# PROJECTS IN SPACE SCIENCE

**OTHER BOOKS BY ROBERT GARDNER**

*Kitchen Chemistry*

*Science around the House*

*Water: The Life Sustaining Resource*

*The Whale Watchers' Guide*

*The Young Athlete's Manual*

**WITH DAVID WEBSTER**

*Science in Your Backyard*

# PROJECTS IN SPACE SCIENCE

ROBERT GARDNER

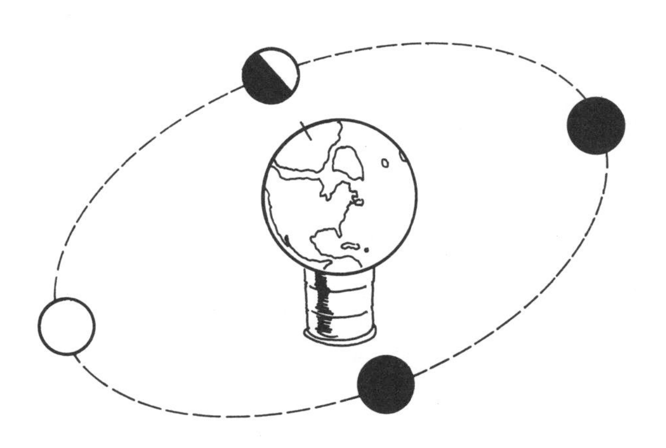

JULIAN MESSNER
NEW YORK
A DIVISION OF SIMON & SCHUSTER, INC.

Photos courtesy NASA
Star maps, reproduced with
permission from *Star Maps for
Beginners,* I.M. Levitt and
Roy K. Marshall, Simon &
Schuster, 1964


Published by Julian Messner, A
Division of Simon & Schuster,
Inc., Simon & Schuster
Building, Rockefeller Center,
1230 Avenue of the Americas,
New York, NY 10020.

JULIAN MESSNER and
colophon are trademarks
of Simon & Schuster, Inc.

Designed by Raul Rodriguez
Manufactured in the United
States of America

10 9 8 7 6 5 4 3 2 (lib. ed.)
10 9 8 7 6 5 4 (pbk. ed.)

Library of Congress Cataloging-
in-Publication Data

Gardner, Robert, 1929-
Projects in space science.

Includes index.
Summary: Experiments
relating to the origin of the
solar system, the laws of
motion, natural forces affecting
orbiting objects, man's future
in space, and other aspects of
space science.
1. Space sciences —
Experiments — Juvenile
literature.
[1. Space sciences —
Experiments. 2. Experiments]
I. Title.
QB500.264.G37 1988
500.2 87-20275
ISBN 0-671-63639-1 (lib. ed.)
0-671-65993-6 (pbk. ed.)

# CONTENTS

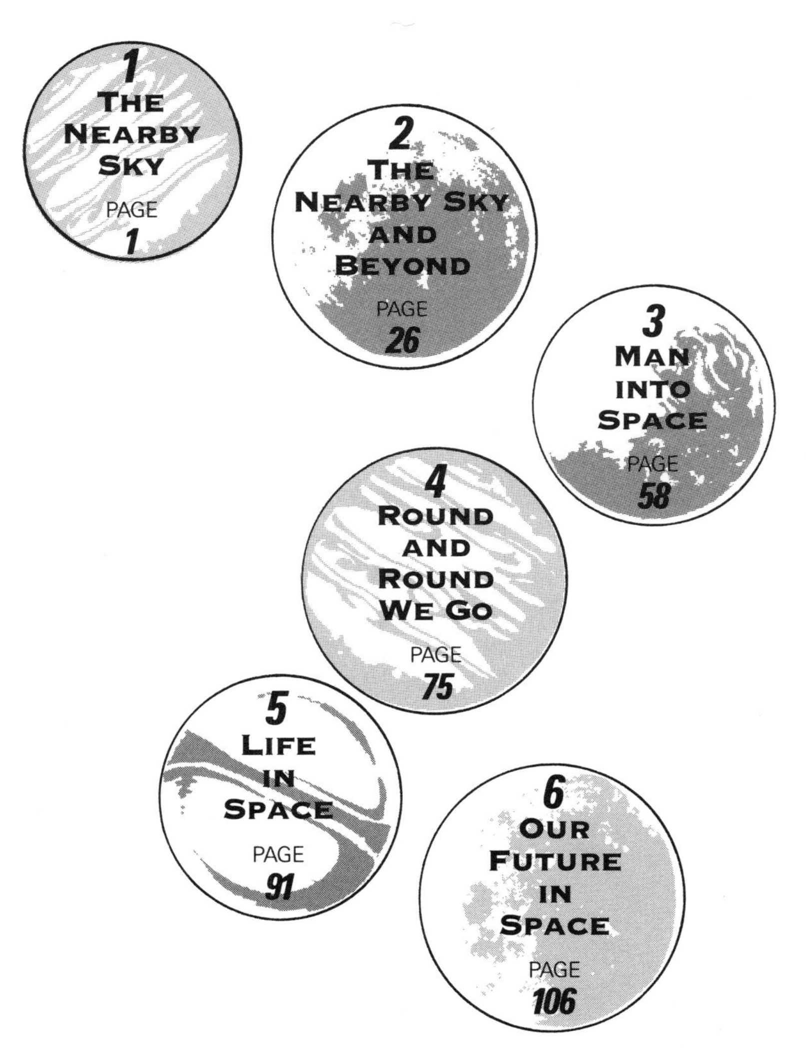
1
THE NEARBY SKY
PAGE 1
2
THE NEARBY SKY AND BEYOND
PAGE 26
3
MAN INTO SPACE
PAGE 58
4
ROUND AND ROUND WE GO
PAGE 75
5
LIFE IN SPACE
PAGE 91
6
OUR FUTURE IN SPACE
PAGE 106

# CHAPTER 1

# THE NEARBY SKY

If you step outside on a clear day and look up, you will see a blue dome we call the sky. (**Never look directly at the sun. It can cause permanent damage to your eyes!**) The sun, sometimes the moon, and, if you look very carefully, the planet Venus can be seen on the surface of this dome, known to astronomers as the *celestial sphere.* At night, you can see thousands of stars, sometimes the moon, and often one or more planets. Most people see stars scattered across the sky in a random way. But others see definite patterns in the stars.

The sun, which is the brightest star we see, rises and sets every day, giving us day and night. People used to think that the sun traveled a circular path about the earth each day. But Copernicus argued that the earth actually turns about its axis each day, which makes the sun *appear* to move around the earth. You're probably convinced of the truth of that explanation, but do you have good experimental evidence to support your belief?

## WHERE ARE WE?

Positions on Earth are established from a giant imaginary grid covering the earth's surface. These are the lines you see on maps or a globe. The lines that run north-south are called *meridians.* These lines measure *longitude.* The prime meridian is 0 degrees longitude. It runs from pole to pole through Greenwich, England. If you look on a globe, you will see that the distance between these longitude lines is greatest at the equator. The lines join to form a point at each pole.

The sun seems to move in a circle about the earth once every 24 hours. Since there are 360 degrees in a circle, the sun appears to move 15 degrees of longitude every hour. That's why time zones are about 15 degrees apart. When you travel westward from one time zone to the next, you must set your clock back one hour. Why aren't all time zones exactly 15 degrees apart?

Imaginary lines parallel to the equator are called *parallels.* They are used to measure latitude — degrees north and south of the equator. Degrees of latitude are about 69 miles (111 kilometers) apart. The North Pole is 90 degrees latitude. The equator is 0 degrees latitude. This book is being written at 42 degrees latitude.

## YOUR LATITUDE

Do you know the latitude where you live? You can make a pretty good estimate of your latitude by measuring the altitude of the North Star (Polaris), which is located almost directly above the earth's North Pole. As you can see in the drawing, the altitude — the angle of elevation above the horizon — of Polaris is equal to the

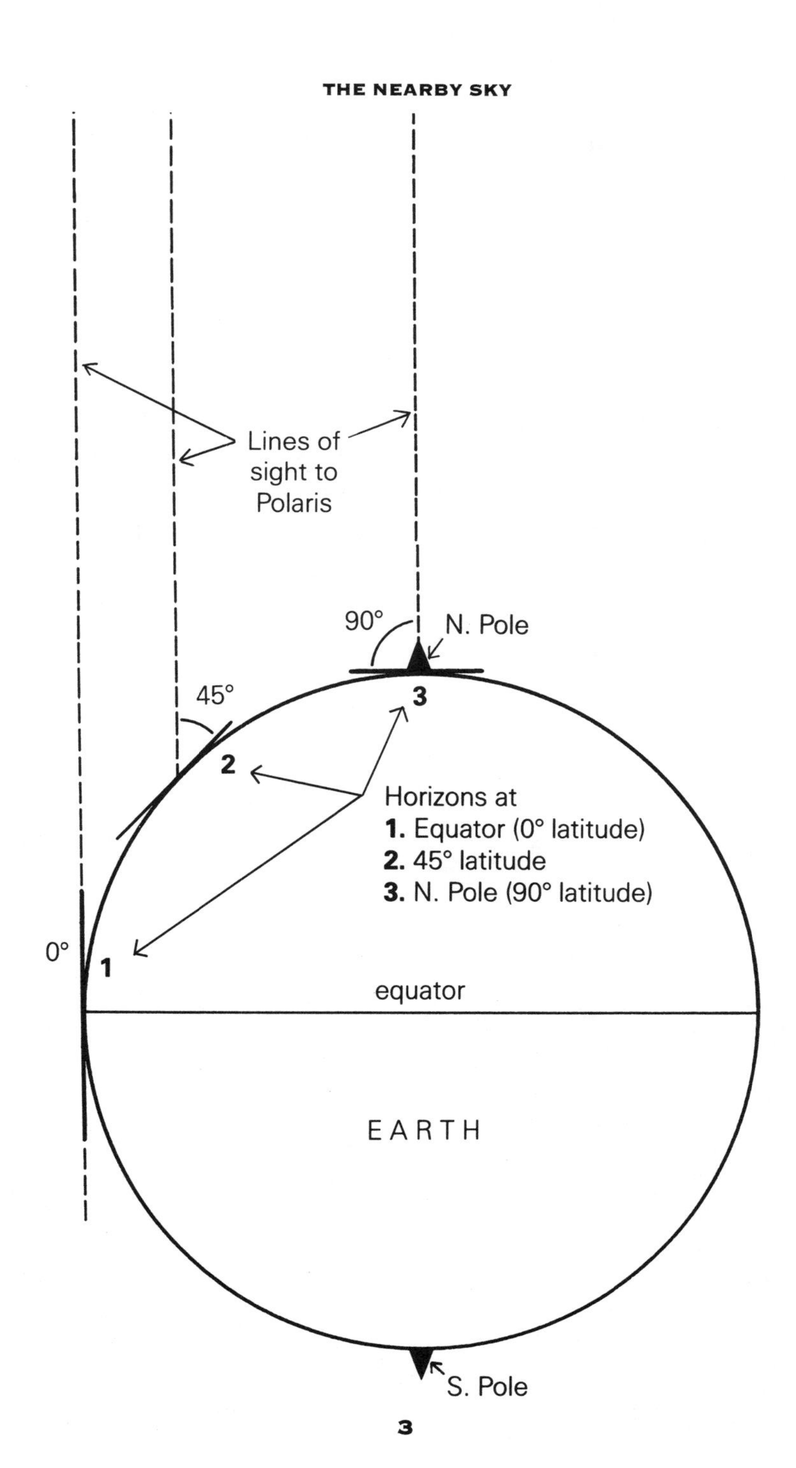
Lines of sight to Polaris
90°
N. Pole
45°
3
2
Horizons at
1. Equator (0° latitude)
2. 45° latitude
3. N. Pole (90° latitude)
0°
1
equator
EARTH
S. Pole

latitude from which its altitude is measured. Because Polaris is so far from Earth, light rays coming from the star are very nearly parallel. Any that were not parallel would not strike Earth.

To find Polaris, go outside on a clear night. Look to the northern half of the sky and find the Big Dipper. It consists of a group of bright stars that looks like the side view of a cooking pan or water dipper, as you see in the drawing. Depending on the season and the time of night, the Big Dipper may be turned at different angles in the sky. (See p. 36 in Chapter 2.) The pointer stars, Dubhe and Merak, form a line that points toward Polaris. The distance of Polaris from the Big Dipper is about five times the distance between these two pointer stars. Do not expect to find a very bright star. Polaris is about as bright as Merak. It is the star at the end of the handle of the Little Dipper.

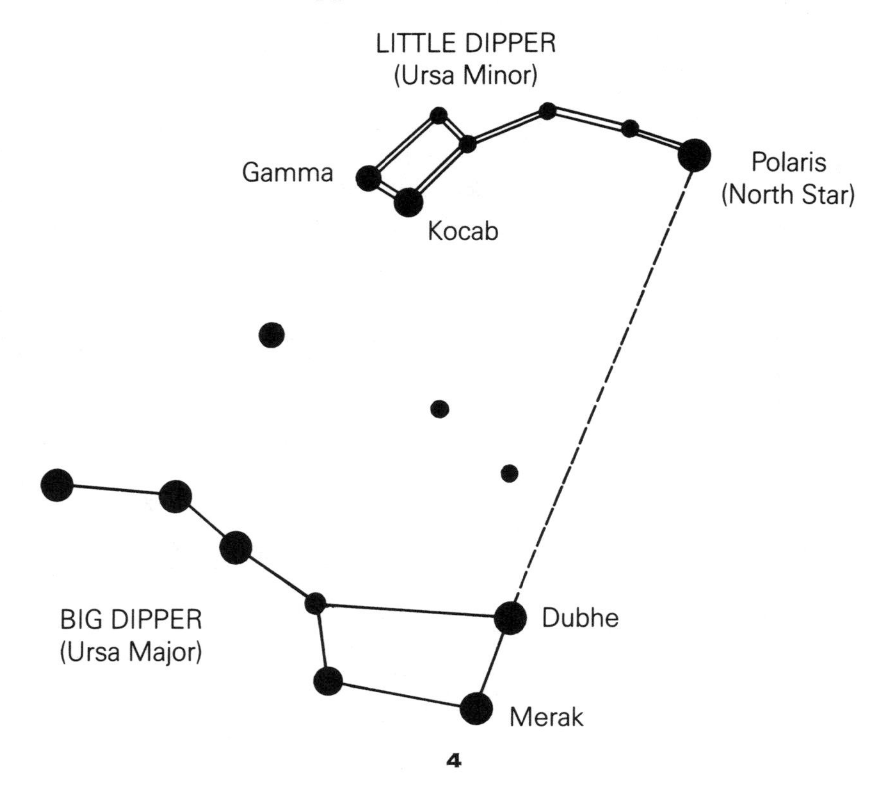

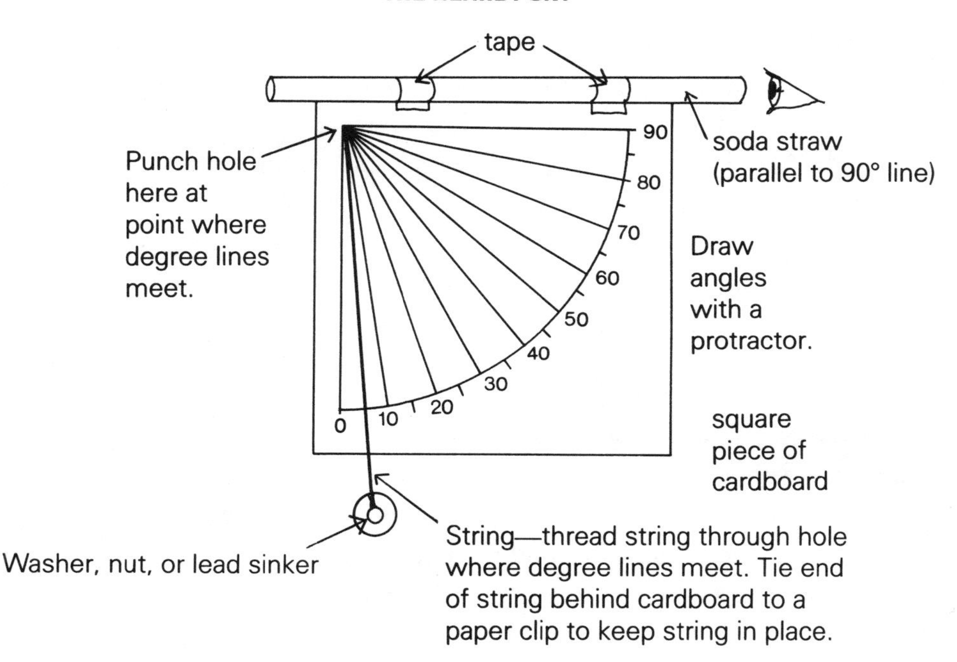


An astrolabe is turned upward; the string marks the altitude in degrees. If the star sighted is overhead, the string will lie on the 90 degree line.

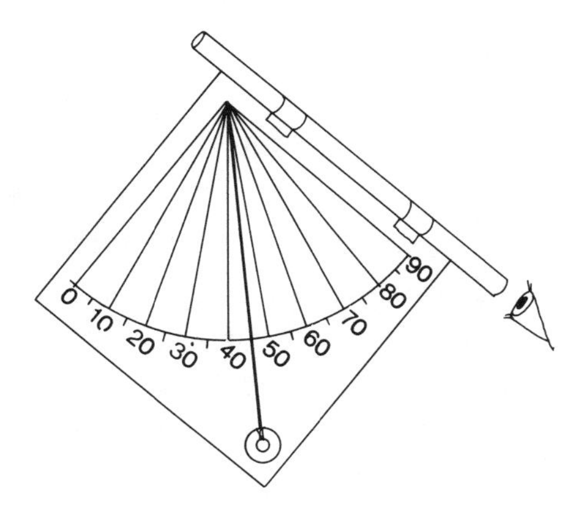


To measure the altitude of Polaris, you can build an astrolabe like the one shown. When you look at the North Star through the soda straw, the string will hang along a line that measures the star's altitude. What is the altitude of Polaris? What is the latitude of your location?

While you are measuring the North Star's altitude, have someone help you mark a north-south line of sight along level ground. While you stand looking at Polaris, your helper can establish a straight line of sight from the star to you to himself or herself. Markers at your position and the position of your helper will provide a north-south line that you will find useful in later experiments.

## The Shifting Sun

The sun does not follow the same path across the sky each day. The summer sun follows a longer and higher path than does the winter sun. To find the path of the sun at various times of the year, you can build a sundial. You will need a square board about 12 inches (30 cm) on a side, a new, unsharpened pencil, white paper, a ruler, and a hand drill.

With the hand drill, make a hole in the board that has the same diameter as the pencil. The hole should be near the center of the length of one side of the board and about 2 to 3 inches in from the edge, as shown in the drawing. Break off the eraser end of the pencil and sharpen it. Then, to prevent injury, break off the lead and sandpaper the end into a smoothly rounded surface. Insert the unsharpened end of the pencil into the hole in the board. The pencil should rise about 4 to 5 inches (10 – 13 cm) above the surface of the board. Place the sundial on a level spot along a north-south line, such as the one you established while measuring the altitude of Polaris. The board should be turned so the pencil, which is called the *gnomon,* is at the south end of the sundial. Use a small carpenter's level to make certain the board is level and the pencil perpendicular to its surface.

Cover the board with a sheet of white paper or brown wrapping paper. Cut a slit along the south edge of the paper so it will fit

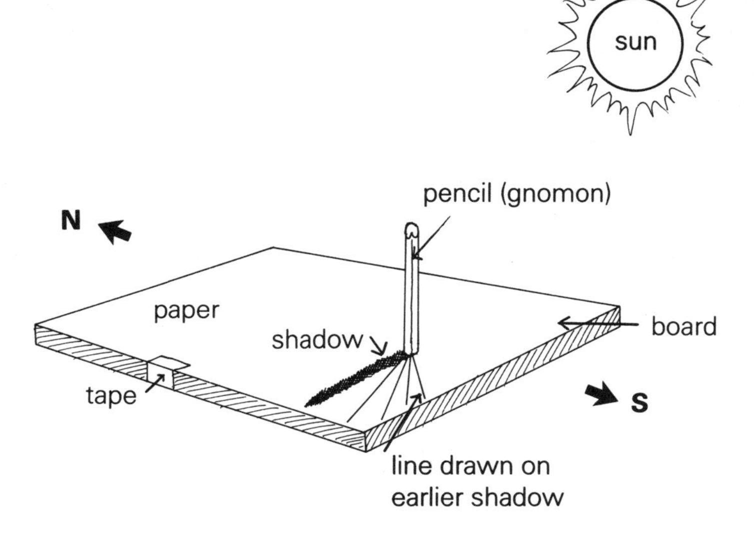


around the pencil, and tape the paper to the board. You are now ready to take measurements.

Early in the morning, when the sun begins to cast a shadow of the pencil on the paper, you can begin taking measurements. With a ruler, draw a line along the center of the shadow from the pencil to the end of its shadow. Measure the line you have drawn. (If the shadow should extend beyond the board, place another board beside your sundial and measure the total length of the shadow.) Write the length of the shadow and the time you took the measurement. Repeat this procedure at frequent intervals throughout the day.

To check up on the north-south line you made, take frequent measurements of the pencil's shadow around midday. The shortest shadow cast by the pencil will occur when the sun is due south. At that time, the pencil's shadow will lie along a north-south line. Is the pencil's shortest shadow parallel to the north-south line you established earlier?

After sunset, you can bring the paper inside and determine the sun's position in the sky at various times during the day. Draw a north-south line (through the shortest shadow) and an east-west line on your paper. Each line should pass through the point where the pencil was located.

The sun's *azimuth* is its angle along the horizon relative to north. North is 0 degrees, east is 90 degrees, south is 180 degrees, and west is 270 degrees. With a protractor, determine the sun's azimuth for each line you drew.

The sun's *altitude* is its angle of elevation above the horizon. If the sun is on the horizon, its altitude is 0 degrees. If it is directly overhead, its altitude is 90 degrees. To determine the sun's altitude at each of the times you took a measurement with your sundial, draw a vertical line equal to the height of the pencil. At the base of this line, draw a line equal to the length of the longest shadow. (This line should make a right angle with the line representing the height of the pencil.) Mark the length of the shadow for each of the times you measured it on this line. A line connecting the shadow's length and the pencil's height will enable you to find the sun's altitude for each of the times you marked the pencil's shadow. Just lay a protractor on the angle, as shown.

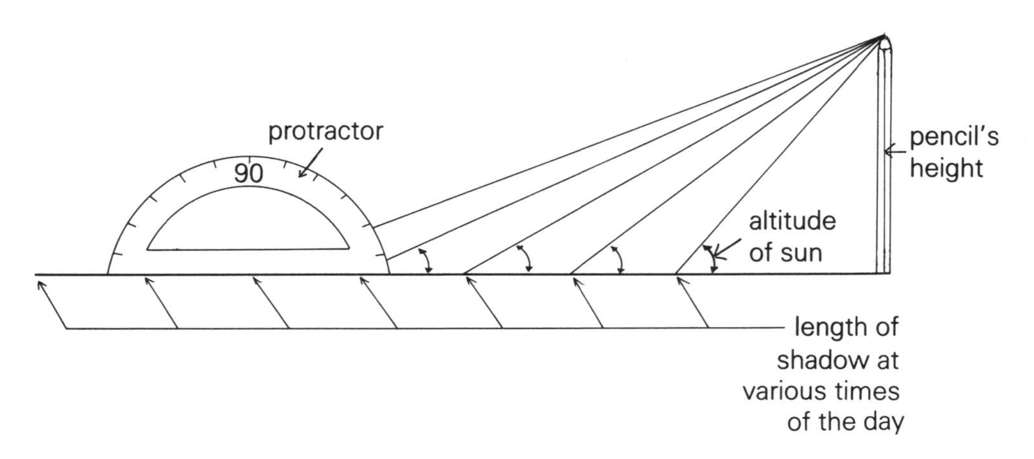

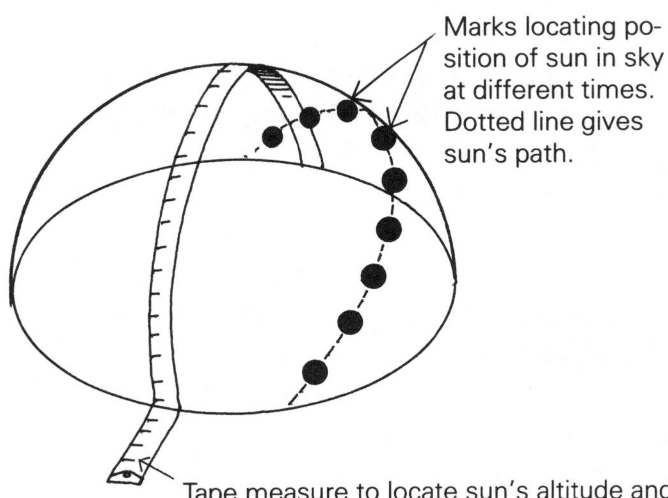


Tape measure to locate sun's altitude and azimuth. If circumference of hemisphere (½ circumference of sphere) is 12 inches, each inch is equivalent to 1⁄12 of 180°, or 15°. Similar measurements can be made to establish azimuth position of sun on the hemisphere.

With the information you have, you can map the sun's position in the sky at each of the times you took a measurement with the sundial. (See the drawing.) You could use map pins in a wire mesh strainer or colored pen marks on a clear dome. By saving this map of the sun's path across the sky, you can compare it with other maps that you can make at different times of the year. To see the biggest changes in the sun's path, take measurements at the beginning of each season — around the 20th of June, September, December, and March.

When is the sun's path across the sky longest? Shortest? Does the sun always rise at an azimuth of 90 degrees and set at 270 degrees? When does the sun reach its greatest altitude?

## A Model to Explain Seasonal Changes

You know that the sun's path across the sky changes from season to season. To explain seasonal changes on Earth, it is important to know that the earth's axis is tilted 23.5 degrees from the perpendicular of its orbit about the sun. To see how this affects sunlight falling on Earth, place a bright light bulb in the center of a table in a dark room. Then move a small globe or a ball in a large circle around the bulb. Keep the ball or globe tipped at the same angle (see the drawing) as you move it along its circular path. Stop at point *S* and turn the ball or globe to represent the earth's rotation on its axis. In which part of the earth does the sun never set when it is at point *S*? In which part does the sun never rise?

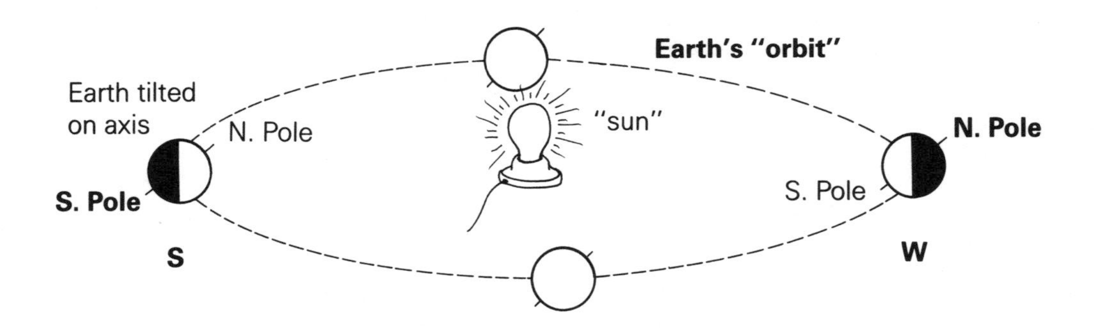


Repeat this process at point *W*. What seasons are represented by this model when the earth is at points *S* and *W*?

To see how the angle at which light strikes the earth affects a season's average temperature, tape a long section of a cardboard mailing tube to the end of a flashlight. Place the end of the tube several inches above and perpendicular to a white sheet of paper, as

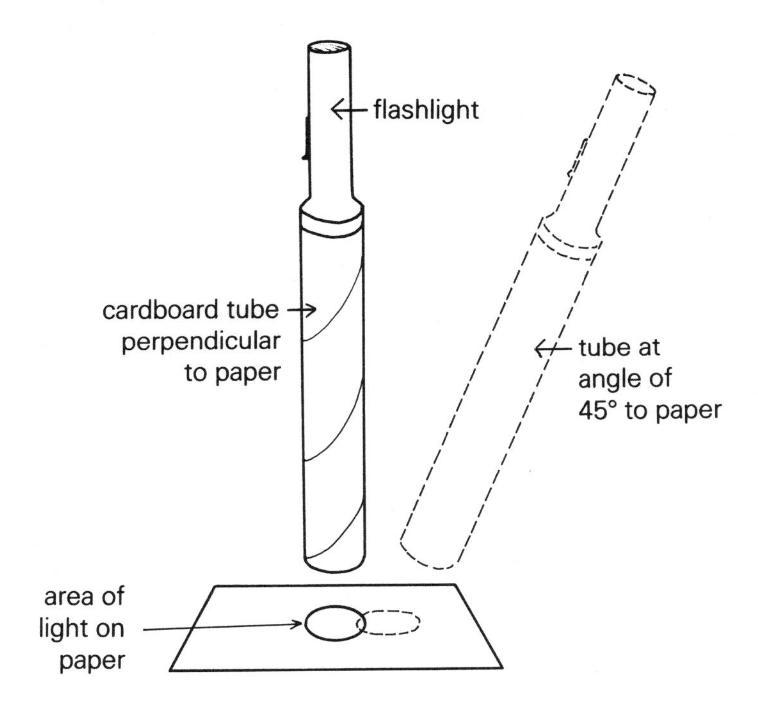


shown. Have a partner mark the outline of the light that falls on the paper. Keeping the tube the same distance from the paper, tilt the flashlight and tube so that the light falls on the paper at a different angle. Again, have someone mark the outline of the light on the paper. Continue changing the angle that the tube makes with the paper until the tube is almost parallel with the paper.

The amount of light coming through the tube is constant, just as light emitted by the sun is constant. But what happens to the area of the light on the paper as the tube moves from a position perpendicular to the paper to one that is nearly parallel? Which condition is closest to summertime? To wintertime? How is the area over which the light's energy spreads related to the season's average temperature?

## THE SUN'S DIAMETER

Earth is about 93 million miles from the sun. With that information, a yardstick, a pin, an index card, and a cardboard screen, you can make a good estimate of the diameter of the sun.

Hold a sheet of cardboard to which you have taped a piece of white paper at one end of a yardstick. At the other end (or closer), hold an index card in the center of which you have made a small hole with a pin. Light from the sun that passes through the pinhole will produce an image of the sun on the screen up to 36 inches away. (**Never look directly at the sun. It can damage your eyes permanently!**) Measure the diameter of the sun's image on the screen. As you can see in the drawing, light rays from the sun that pass through the pinhole to form the image make two similar triangles. This means that the ratio of the sun's diameter to its distance from the pinhole (93,000,000 miles) is the same as the ratio of the diameter of the image to the length of the yardstick (36 inches).

Since you can calculate the ratio of the image's diameter to 36 inches, you can find the diameter of the sun, in miles, that gives the same ratio when divided by 93,000,000.

If you have a telescope or binoculars, you can make larger images of the sun. *Do not* under any circumstances use the telescope or binoculars to look directly at the sun! **It is particularly dangerous to look directly at the sun through any magnifying device such as a telescope, binoculars, or magnifying glass.** Use a telescope to project the sun's image onto a screen such as a sheet of cardboard. You'll have to experiment to find the lens adjustment and screen distance that gives the largest and clearest image.

If you are able to produce such an image, you may find small, dark blotches on it. These are sunspots. Watch them over the course of several days to see how they move. Sunspots are regions of the sun's surface where the temperature is more than a thousand

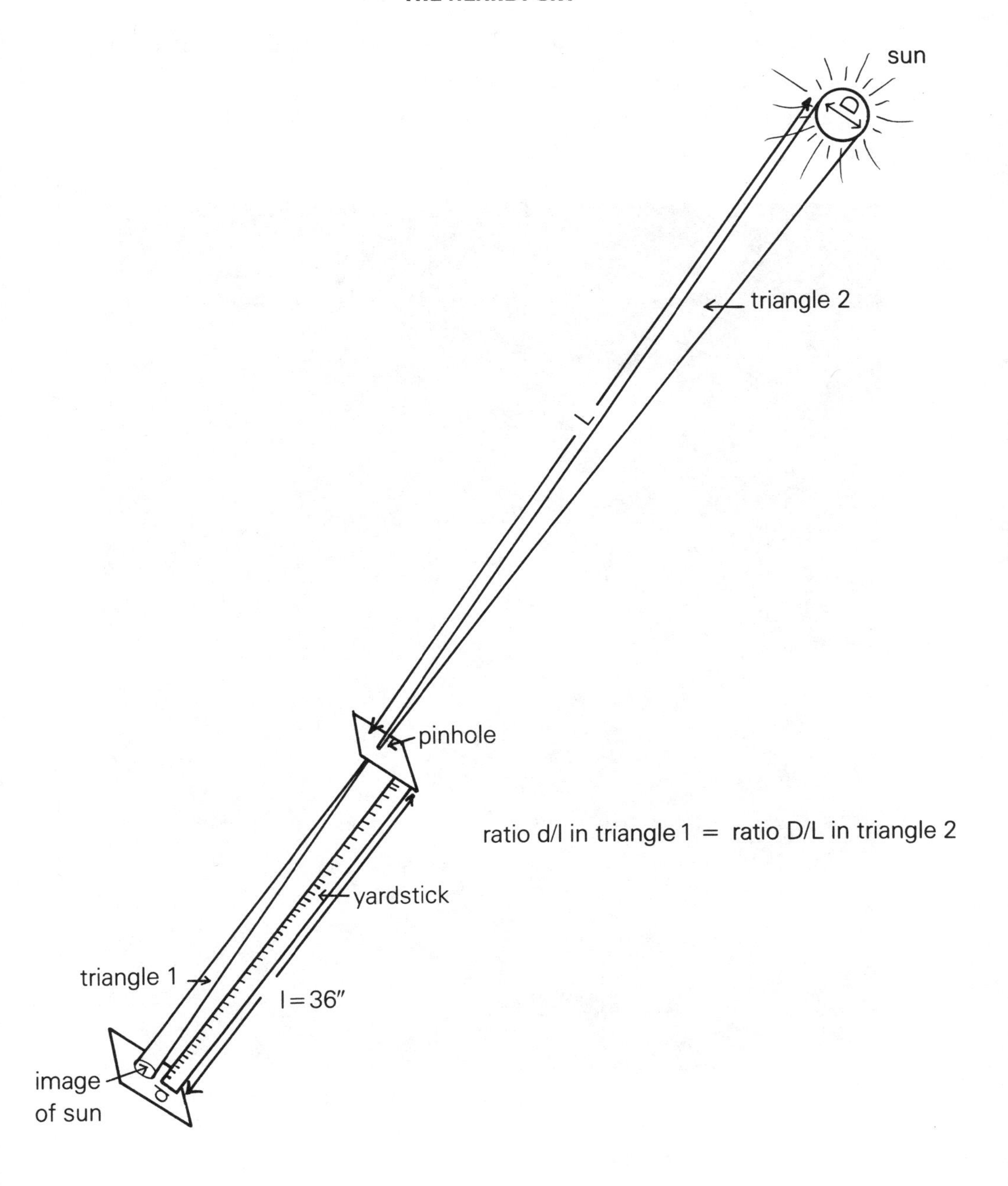

sun
D
triangle 2
L
pinhole
ratio d/l in triangle 1 = ratio D/L in triangle 2
yardstick
triangle 1
l = 36″
image
of sun
d

degrees cooler than elsewhere. They are believed to be caused by intense magnetic fields within the sun.

The number of sunspots reaches a maximum about every 11 years. The next maximum is expected around 1991.

**An eclipse of the sun as seen from aboard Apollo 12. As you can see, the moon appears to have very nearly the same size as the sun. How can this fact be used to determine the distance to the sun?**

## SOLAR ENERGY

If you stand in sunlight while wearing dark clothing, you will be aware that sunlight carries energy because you will feel the warmth. To see how much solar energy is reaching your location on Earth, put about 50 milliliters (ml) of water that is several degrees cooler than the outside air into a Styrofoam cup. Add a few drops of black ink, red, blue, and green food coloring to the water until the liquid is very dark. To insure good insulation, place the cup in an identical empty cup.

Record the water temperature in the shade and then place the water in sunlight. Turn and tip the cup so that the sun's rays shine directly into the cup. (You'll know you've found the right position when there are no shadows in the cup.) Record the time that you turn the cup toward the sun. Leave the cup in the light until the temperature, after stirring, rises about as far above air temperature as it was below when you started the experiment. Then return the cup to a shady place and record how much time has elapsed. How does raising the temperature of the water as far above air temperature as it was below help to reduce errors that might arise because of heat losses or gains from sources other than sunlight? What was the change in temperature? How long did you keep the water in sunlight?

To measure the area of the water exposed to sunlight, mark the cup at the height of the water level. Use scissors to cut off the top of the cup at this height, then measure the resulting diameter. The area of the water surface at this height will equal the square of the diameter of the cup, in centimeters, times pi (3.14), divided by 4. (Area = $\pi d^2/4$.) What was the area of the water surface on which sunlight fell?

A calorie is the amount of heat needed to raise the temperature of 1 gram of water 1 degree Celsius (1.8°F). Since 1 milliliter of water weighs 1 gram, you had 50 grams of water. How much heat was transferred to the water by the sunlight? For example, if the temperature rose 4 degrees Celsius, the heat delivered was 4°C x 50 g = 200 calories.

Measurements of the sun's energy above the earth's atmosphere show that about 2 calories of energy fall on each square centimeter every minute. How much of that energy per square centimeter per minute is available at your location according to your measurements? Does it vary from day to day? What weather conditions allow the greatest amount of solar energy to come through the atmosphere?

## The Diameter of the Moon

The moon is only about 240,000 miles away, so it is much closer to us than the sun. You can use similar triangles to measure the diameter of the moon just as you did for the sun. However, the moon is not bright enough to produce pinhole images, so you'll need a different method than the one you used to find the sun's diameter. Cut a square, one-half inch on a side, in an index card. Have someone move the card away from your eye until the diameter of the moon fills the half-inch height of the square hole in the card. Now you have the two triangles shown in the drawing. What do you estimate the diameter of the moon to be based on your measurements?

If you look at the moon as it rises, it appears much larger than it does when it is higher in the sky. Is this an illusion? How can you find out?

## Moon Changes

Like the sun, the moon seems to circle the earth. But if you observe moonrise daily, you'll see that it changes from one day to the next.

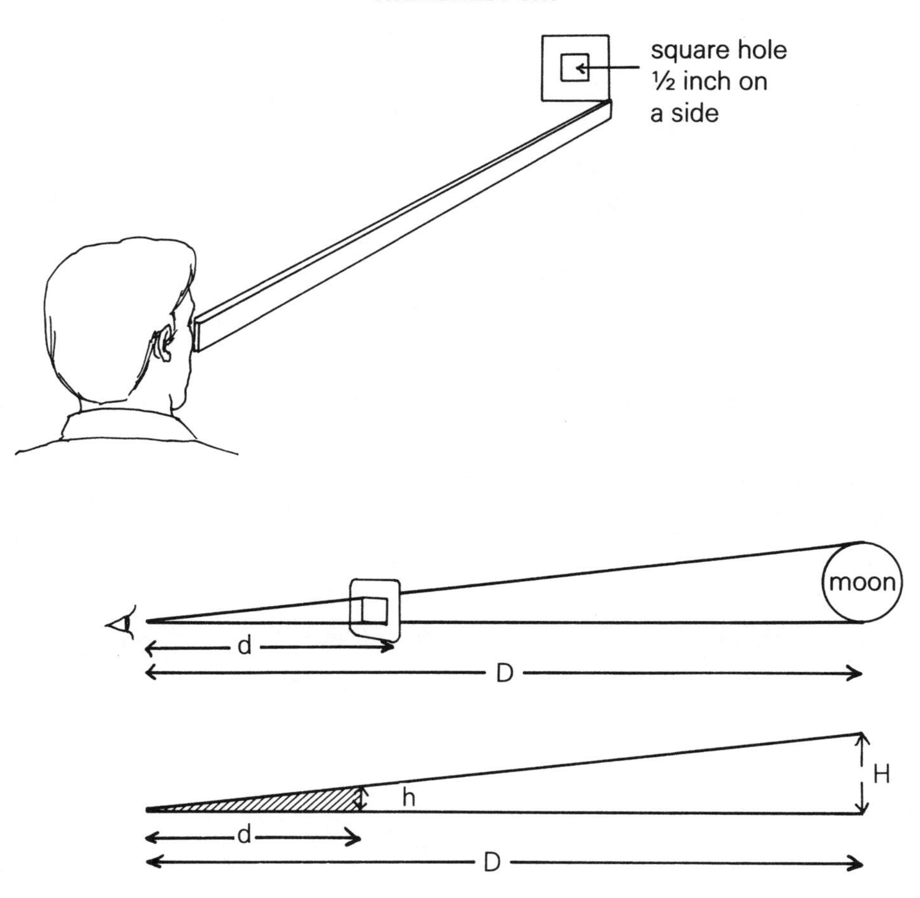


The little shaded triangle from eye to square hole is part of the big triangle from eye to moon. Because the little and big triangles are similar, the ratio d/h = the ratio D/H.

You'll notice, too, that the apparent shape of the moon seems to change dramatically over the course of a few days.

Try to observe the moon over a period of several months. Make a daily sketch of its shape — when you can see it — a record of where it rises (its azimuth) and/or sets, and its altitude and direction at different times. Do this as often as possible. You can make rough estimates of azimuth and altitude using your fists. If you hold your fists at arms length from your body, each fist occupies about 10 degrees. Try it! Hold your left fist so the top of your hand is level with the horizon. Then put your right fist on your left, then your left on your right, etc. You'll find that it takes just about nine fists to reach a point directly overhead (90 degrees above the horizon).

How does the position of a rising full moon compare with the position of the setting sun on the same day? See if you can learn to predict the time, location, and shape of the rising moon. How does the path of the moon across the sky compare with the sun's path?

You can make a model of the moon's path about Earth that will help you understand why the moon's shape changes. Place a globe, representing the earth, outside on a sunny day. Have a friend move a tennis ball about the globe, as shown in the drawing, while you look at the ball with your eyes near the globe. Where, relative to Earth and sun, does the ball represent a full moon as seen from Earth? A first quarter? A last quarter? A new moon?

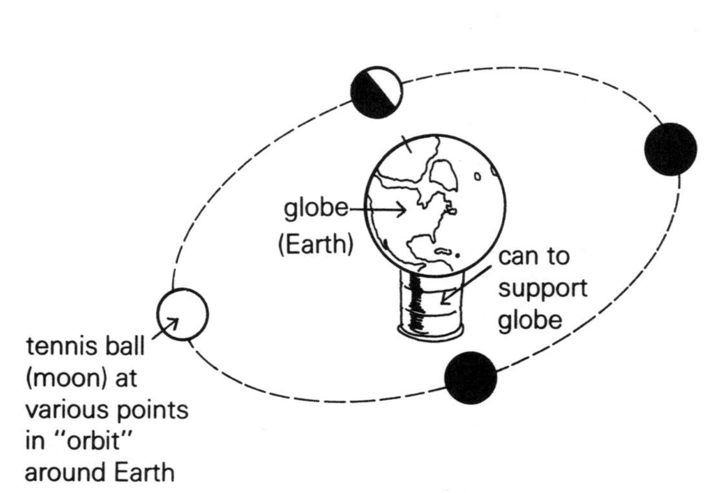

Using the model, when would you see a "full earth" if you were on the moon?

When the earth comes between the moon and the sun, the earth's shadow falls on the moon, and we see an eclipse of the moon. Where should the ball be placed to represent an eclipse of the moon?

If the moon comes between the earth and the sun, it casts its shadow on the earth, causing an eclipse of the sun. Where should the ball be placed to represent an eclipse of the sun? **Do not look at the sun!**

Make a scale model of the earth and moon from what you know about their diameters and separation. Attach the "earth" to one end of a long stick that represents the distance of earth to moon on your scale. Attach the moon to the other end. Take this model out into sunlight and move it about to produce eclipses. Why do eclipses occur so infrequently?

## Other Views of the Moon

To see the moon when it is a crescent, after a new moon, you'll have to view it shortly after sunset. You may be able to see a faint outline of the rest of the moon as well as the bright crescent. This is sometimes called "the old moon with the new moon on its arms." The light beyond the crescent is due to earthshine—sunlight reflected from the earth to the moon and then back to you.

If you have binoculars or a telescope, you'll enjoy seeing a magnified image of the moon and perhaps sense the same thrill that Galileo felt when he became the first human to view the lunar surface through a telescope. If you look along the terminator, the edge of darkness on the moon, you'll see that the long shadows cast there make the craters more distinct. Keep viewing the moon from one new moon to the next. You'll notice that you always see the

**A full moon as seen from Apollo 11 during its homeward journey in 1969.**

same side of the moon. What does this tell you about the moon's period of rotation (the time it takes to turn once on its axis) as compared with its period of revolution (time to make one complete orbit around the earth)?

## A Look at Some Planets

After the moon, the brightest object in the sky is Venus. It is often called the morning or evening "star." Of course, it's not a star. All the light we see from Venus, or any of the planets, is sunlight that is reflected from its surface. Because Venus is covered with clouds, 75 percent of the sunlight that strikes the planet is reflected. (Only 40 percent of the light striking Earth is reflected.) Since Venus's orbit is closer to the sun than ours, Venus never appears very far from the sun. Your newspaper will probably tell you the rising and setting times for Venus, so you should have no trouble finding it in the morning, before sunrise, or in the evening, after sunset, unless it happens to be very close to the sun. Watch Venus for several months. What's the largest angle it makes with the sun? You can use your fists to estimate this angle as the sun sets or rises. Remember: **Never look at the sun!**

If you locate Venus just before sunrise and keep track of its position, you'll be able to see it even after the sun rises. It's fun to point it out to others and show them that a "star" can be seen even in the daytime.

Again, with the newspaper to help you, you can probably locate the planets Jupiter, which is quite bright, and Mars, which is very red. Mercury, which is never more than 28 degrees from the sun, and Saturn, the bright, ringed planet, are more difficult to find, but with patience and perseverance you will see them. Binoculars will help, but if you're looking for Mercury **don't look while the sun is in the sky.** There are a number of moons that orbit Jupiter, and you can

**Six planets of our solar system. Photographs of six planets were taken at different times from NASA spacecraft. In the foreground, the Earth is seen rising over the moon's surface. A sun flare is seen at the edge of the Earth. The planets seen from left to right in the background are Jupiter, Venus, the left side of Mercury, the top of Mars, and Saturn.**

see four of them with binoculars; however, you will probably have to mount the binoculars on a tripod or hold them against a firm object.

On a clear, dark night you may see meteors, or shooting stars as they are often called. These are particles of matter that burn up when they hit the earth's atmosphere. There are particular places along the earth's orbit about the sun where these particles seem to be concentrated. When the earth crosses these places, you have an opportunity to see forty or fifty meteors per hour "shower" through the sky. The Perseids shower can best be seen about August 12, and the Geminids shower about December 13. The showers are named for the constellation where they appear in the sky. Most newspapers will have an article about these two meteor showers just before they occur.

## Bode's Law

During the late 1700s, astronomers developed a number scheme that became known as Bode's law. If you write down the series of numbers 0, 3, 6, 12, 24, 48, 96, add 4 to each number, and then divide each by 10, you obtain: 0.4, 0.7, 1.0, 1.6, 2.8, 5.2, 10.0.

You may think these are just a bunch of random numbers, but the numbers were significant to astronomers who often measure distances in astronomical units (AU). One astronomical unit is the distance from the earth to the sun (93 million miles). If a planet were half as far from the sun as the earth is, the radius of its orbit would be 0.5 AU. A planet twice as far from the sun as the earth is would have a radius of 2.0 AU.

Table I lists the planets and the radii of their obits in astronomical units. Notice how closely the radii match the numbers in Bode's law. You can see why astronomers found Bode's law significant.

**TABLE I**

| **Planet** | Mercury | Venus | Earth | Mars | ? | Jupiter | Saturn |
|---|---|---|---|---|---|---|---|
| **Radius (AU)** | 0.39 | 0.72 | 1.0 | 1.52 | 2.8 | 5.2 | 9.54 |

In 1781, the planet Uranus was discovered. Its orbit had a radius of 19.2 AU. This is close to the next number in the Bode's law series: (192 + 4)/10 = 19.6.

The radii of the planets, including Uranus, are close to the numbers in the Bode's law series, but there is no planet at 2.8. During the nineteenth century, astronomers began to find small bodies at about 2.8 AU from the sun, in the region between Mars and Jupiter. These bodies, with diameters of 1000 kilometers or less, are known as asteroids or minor planets. It used to be thought that the asteroids were the remnants of a planet that exploded. But it is more likely that they are particles from the original solar system that never came together to form a planet.

Meteors are believed to be material left by comets that broke up as they moved about the sun. These particles continue to move along the comet's orbit—an orbit that crosses Earth's orbit.

Occasionally, a comet will become visible as it moves close to the sun. The orbits of comets are very long ellipses, unlike the nearly circular orbits of most planets.

A meteorite is a chunk of rock that has survived its fiery path through the atmosphere and struck the earth. Some meteorites have been as large as 27 feet across. The New Quebec Crater, which was created by a meteorite, is nearly 2 miles (3 kilometers) in diameter. Fortunately, large meteorites are very rare.

## LIFE ON EARTH

Though nine planets (Mercury, Venus, Earth, Mars, Jupiter, Saturn, Uranus, Neptune, and Pluto) orbit the sun, only Earth seems to harbor life. Mercury is sun-scorched, the surface of Venus, where it rains sulfuric acid, has a temperature of 870°F (470°C), and the planets from Jupiter outward are very cold. There is evidence of water on Mars, but landings there in 1976 by Vikings I and II revealed no evidence of life. And astronauts found our moon, which has no atmosphere or water, to be lifeless.

On Earth, the evolution of plants that can produce oxygen and food from water and carbon dioxide has provided an oxygen-rich atmosphere where animal life can prosper.

About thirty-five years ago, Harold Urey and Stanley Miller mixed water vapor, methane, and ammonia in a flask. These gases are thought to have been present in the earth's early atmosphere. Electric sparks, similar to the lightning flashes that were probably prevalent in the earth's early days, were sent through these gases for a number of weeks. At the conclusion of the experiment, the flask contained a variety of organic substances including amino acids, the building blocks of proteins. We still do not know how these substances gave rise to life. But we do know that the basic chemical ingredients of life could have been produced in the earth's primitive atmosphere.

We know also that about 63 million years ago 65 percent of the existing species on Earth, including the dinosaurs, became extinct. Recently, scientists have found a thin layer of iridium-rich dust in sites that are about 63 million years old. Since iridium is an element commonly found in asteroids and meteoroids, some scientists believe that a giant meteorite may have slammed into the earth at that time, spewing huge clouds of dust into the atmosphere. The dust so reduced the sunlight reaching the earth that temperatures fell below levels necessary for the survival of many plants and animals. Following this long winter, the dust gradually settled, allowing sunlight to fall in abundance again on an earth now devoid of many of its previous life forms.

# CHAPTER 2

# THE NEARBY SKY AND BEYOND

The nine planets and the asteroids that circle our sun make up our solar system. But what is the origin of the solar system and the stars that we see beyond it?

It is believed that gravity pulls together the stardust from earlier stars that blew apart to form new planets and stars. Gravity is the force of attraction that one piece of matter exerts on another. This force depends on the amount of matter, commonly called mass, in each piece and the distance between the masses. If the mass of either piece is doubled, the pull between them (force of attraction) will double. If the distance between the pieces doubles, the force will be reduced to one-fourth its former value.

Earth pulls on you. The force it exerts on you is called your weight. If a person has twice your mass, the gravitational force between that person and Earth will be twice as great as the force between you and Earth. He or she will weigh twice as much as you do. But if you were to double your distance from the center of the earth, the earth's pull on you at this greater distance would be much less. Your weight would be quartered.

## ORIGIN OF THE SOLAR SYSTEM

Originally, the rotating stardust that formed our solar system was very cold. The only gases that could exist at such low temperatures were hydrogen and helium. All other matter was probably in the form of solid dustlike particles. Gravitational pulls among the particles or a shock wave from a nearby supernova explosion caused a general drift toward the center of the "cloud," producing a contraction of the matter and higher temperatures. Because of rotation, the cloud became a flattened disk. Near the center, where temperatures were higher because of greater pressure, a protosun formed. Rotation kept some of the matter far from the center—out where it was still very cold.

Continued contraction of the central part of the disk raised temperatures to millions of degrees. At such a high temperature, hydrogen began fusing to form helium, like a giant hydrogen bomb. The energy released by fusion kept that reaction going while creating an outward pressure that balanced the inward gravitational contraction. Thus a stable star, our sun, was born.

The small, dense, inner planets, Mercury, Venus, Earth, and Mars, formed during a period of 100 million years as particles of matter circling the sun were pulled together by gravitational forces. These particles were rich in sulfur, silicon, iron, magnesium and other metals from dead stars. The decay of radioactive elements, the violent impact of the colliding particles growing into planets, as well as heat from the growing star (sun) nearby, kept these materials in a melted state. The denser iron sank beneath the lighter matter.

This explains why Earth has an iron core surrounded by less dense rock and silicon-rich sand. The high temperature caused molecules of hydrogen and helium to move so fast that they escaped from the atmospheres of these inner planets.

For each planet there is an escape velocity. Anything moving faster than the escape velocity can escape the planet's gravity. On Earth, the escape velocity is 24,000 mph. Space probes sent from Earth to explore other planets had to be accelerated to the escape velocity in order to leave Earth and not be pulled back by gravity. The velocities of hydrogen and helium molecules, at the temperatures found on the inner planets, were large enough to escape the relatively weak gravity of these small planets.

The outer planets, Jupiter, Saturn, Uranus, and Neptune, are large and have low densities because so much of their matter is gaseous. They coalesced from particles of solid matter to form protoplanets. Temperatures so far from the sun were low enough that the huge amounts of hydrogen and helium swept up by these planets as they moved along their orbits could not escape. As a result, the outer planets have earth-sized cores of solid matter surrounded by thick gaseous atmospheres.

At about the time the planets reached their present size, the onset of fusion reactions at the sun's core caused it to expel its outer matter. This burst of matter spreading across the solar system carried away any remaining gases and so prevented any further growth of the outer planets. An abundance of small rocks remained to put pock marks on the planets over the next half billion years, but the solar system was essentially finished after the sun began fusing hydrogen into helium.

The outermost planet, Pluto, does not have the characteristics of the other outer planets. Though it is so far away and so small that it is difficult to determine its properties very accurately, it appears to be more like an inner than an outer planet. One theory holds that

**Saturn, an outer planet, as photographed from Voyager 2 at a distance of 21 million kilometers (13 million miles). Note three of Saturn's moons at the left of the planet.**

Pluto was once a moon of Neptune along with Neptune's present moons—Triton and Nereid, which have unusual orbits. An unknown planet passing close to Neptune disturbed the orbits of Triton and Nereid. At the same time, it sent Pluto into a separate orbit about the sun while tearing it into its present form, which includes its moon, Charon.

Table 2 contains information about each of the planets in our solar system.

**TABLE 2**

| Planet | Radius of orbit (AU) | Time to orbit Earth (years) | Diameter (Earth = 1) | Mass (Earth = 1) | Density (g/cm³) |
|---|---|---|---|---|---|
| Mercury | 0.39 | 0.24 | 0.38 | 0.06 | 5.4 |
| Venus | 0.72 | 0.62 | 0.95 | 0.82 | 5.2 |
| Earth | 1.0 | 1.0 | 1.0 | 1.0 | 5.5 |
| Mars | 1.52 | 1.88 | 0.53 | 0.11 | 3.9 |
| Jupiter | 5.2 | 11.86 | 11.3 | 318.0 | 1.3 |
| Saturn | 9.5 | 29.46 | 9.44 | 95.2 | 0.7 |
| Uranus | 19.2 | 84.01 | 4.10 | 14.5 | 1.2 |
| Neptune | 30.1 | 164.8 | 3.88 | 17.2 | 1.7 |
| Pluto | 39.4 | 247.7 | 0.2? | 0.002? | 1? |

## A Scale Model of the Solar System

You know from earlier experiments and reading that the diameter of the sun is about 865,000 miles, the distance from Earth to sun (the radius of Earth's orbit) is about 93,000,000 miles, and the diameter of the earth is about 8,000 miles. Using this information and the data in Table 2, construct a scale model of the solar system. It will give you a good sense of the vast distances between planets.

## THE MILKY WAY

Stars are not spread evenly through space. They cluster in groups called galaxies. Our own sun is but one of the billions of stars that make up our galaxy. On a clear, moonless night, away from city lights, you can see a hazy band that stretches across the sky. Look at this band with binoculars or a telescope. You will see that it is made up of a vast number of stars. These stars make up our galaxy. When we look at them, we are looking edge on into the Milky Way galaxy, the galaxy in which our own sun lies.

The Milky Way galaxy is about 100,000 light years (ly) in diameter. A light year is the distance that light travels in one year. Since the speed of light is 186,000 miles per second, and since there are about 31 million seconds in a year, a light year is a distance of about 6 trillion miles. Consequently, the distance across our galaxy is about 600,000 trillion miles. Such distances are hard to imagine, especially when you realize that our solar system is *only* about 10 billion miles in diameter. This means that 60 million solar systems like ours could fit into our galaxy. Then, when you realize that there are galaxies out to at least 8 billion light years from us, you begin to appreciate the vastness of the universe we live in.

If an inch were used to represent the diameter of the earth, the solar system, on this scale, would stretch 15 miles. The Milky Way galaxy would extend beyond the sun to Saturn.

Galaxies are far apart, but telescopes reveal that they sometimes merge or collide. Since galaxies are only about 100 diameters (10 million ly) apart and stars within galaxies are about 100 million diameters apart, it's not surprising that galaxies collide more often than stars.

The center of Andromeda Galaxy.

## CONSTELLATIONS

The distance to the nearest star outside our solar system is about 4 light years, but a nearby star may not appear as bright as a star that is much farther away. These stars that make up a constellation, such as the Big Dipper, may differ greatly in their distances from us. Between these bright stars are stars too dim to be seen with the naked eye. The stars that form constellations are not necessarily very close together. And because the stars in them may be moving in different directions at different speeds, the constellations that we see today may look very different a few thousand years from now.

Those new to stargazing should be aware that the same group of stars may not have the same name in different countries. The bright constellation that we see as the Big Dipper is called Charles' Wagon or the Plow in England.

In 1927, astronomers from all over the world agreed on the names and locations for 88 separate constellations. If you become an astronomer, you will refer to the Big Dipper or the Plow as Ursa Major (the big bear) and to the Little Dipper as Ursa Minor.

Because the earth moves about the sun, the stars that we see when darkness falls change from month to month. The sky appears to turn about 30 degrees from one month to the next because the earth moves about 30 degrees along its 360 degree orbit in each of the twelve months. Ursa Major's position at 9 P.M. on October 1 will be the same as its position at 7 P.M. on November 1 or 11 P.M. on September 1. If you view the stars at the same time each night, they will appear to move about two hours (30°) westward from one month to the next.

## A SKY CLOCK

Because the sky seems to rotate about the earth at a steady rate, you can build a sky clock that will allow you to tell time on clear nights.

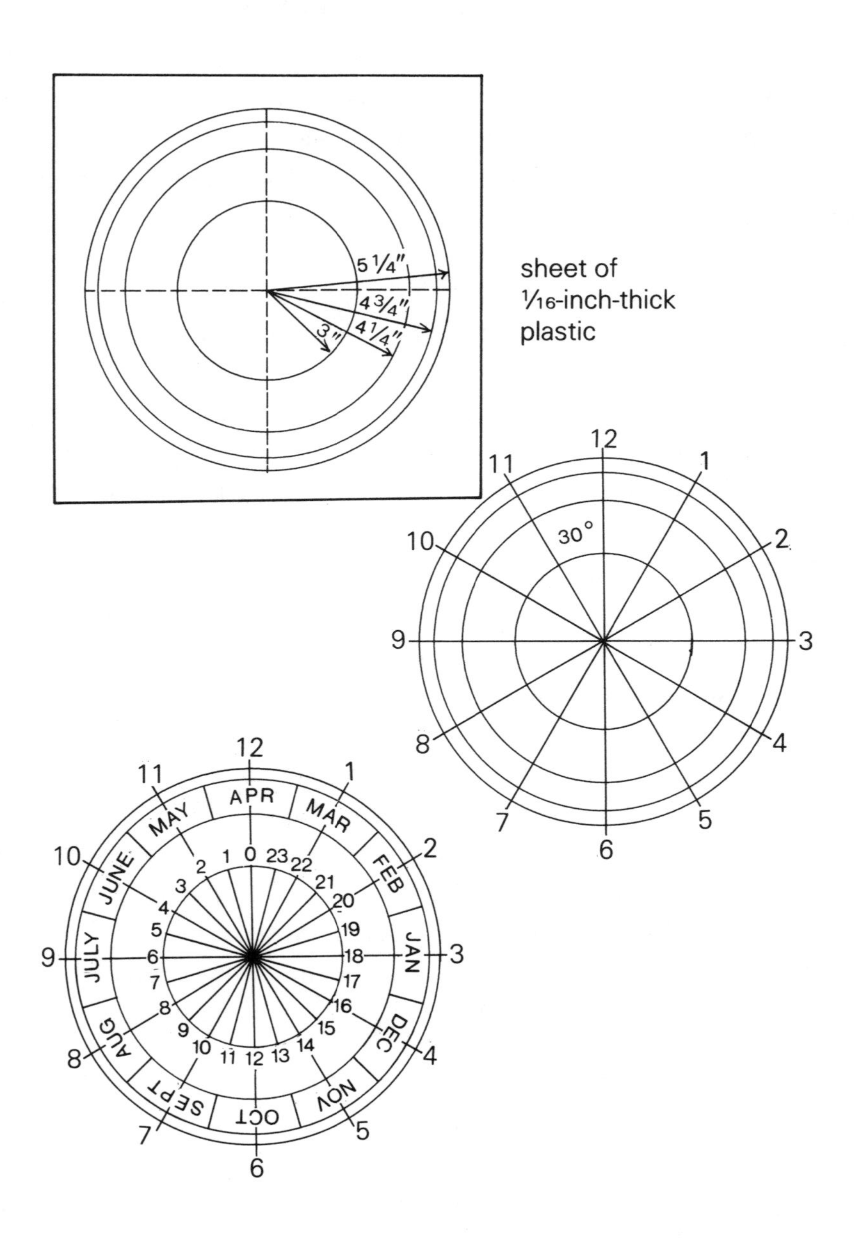
5 1/4"
4 3/4"
4 1/4"
3"
sheet of
1/16-inch-thick
plastic
12
1
2
3
4
5
6
7
8
9
10
11
30°
APR
MAR
FEB
JAN
DEC
NOV
OCT
SEPT
AUG
JULY
JUNE
MAY
0
1
2
3
4
5
6
7
8
9
10
11
12
13
14
15
16
17
18
19
20
21
22
23

To make your sky clock, you'll need a one foot square sheet of 1/16-inch-thick clear plastic. You can buy the plastic in an art store. Draw two straight lines forming right angles through the center of the sheet as shown in the drawing. Place one point of a scriber attached to a compass at the center of the sheet. Use the other point of the scriber to scratch a circle with a radius of 5¼ inches on the plastic sheet. Then draw additional circles with radii of 4¾, 4¼, and 3 inches.

Use a protractor to break the circles into twelve 30 degree segments. With a marking pen, write the numbers found on a clock's face on your sky clock. Now divide each 30 degree segment in half. Use the marking pen to write in the months of the year and the 24 hours represented by the 15 degree lines, and to color the lines scratched in the plastic. The numbers representing hours on the 24-hour inner circle are numbered in counterclockwise fashion because the earth turns counterclockwise as viewed from Polaris. Another, less expensive way to make a sky clock is to stretch a sheet of plastic wrap over a one foot square hole cut from a sheet of heavy cardboard. Tape the sheet firmly and draw the lines on the sheet with a felt pen.

Take your sky clock outside in the evening and early morning. Practice using it to tell time. Hold the clock so that as you look through it its plane is perpendicular to, and its center in line with, Polaris. The 12 directly above April should be at the top of the clock as would be the 12 on a clock in your house. Read the position of the pointer stars of the Big Dipper on your sky clock. Together they form a line representing the hour hand of a clock.

The drawings on the next page show the 9 P.M. positions of the pointer stars at mid-month for each month of the year. Using those drawings, and the position of the pointer stars on your sky clock, you can make a good estimate of the time. For example, suppose it's mid-July. When you hold your sky clock properly, you see the

JANUARY
(3 o'clock at 9 P.M.)

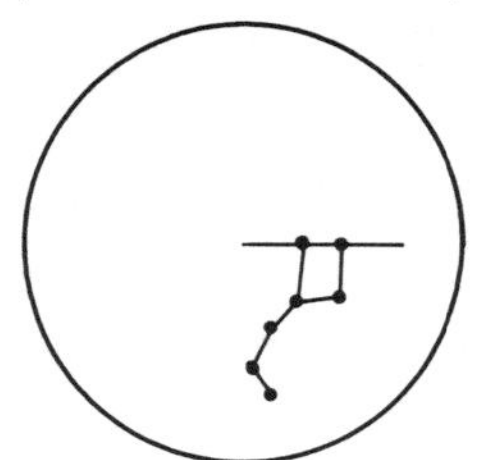

FEBRUARY
(2 o'clock at 9 P.M.)

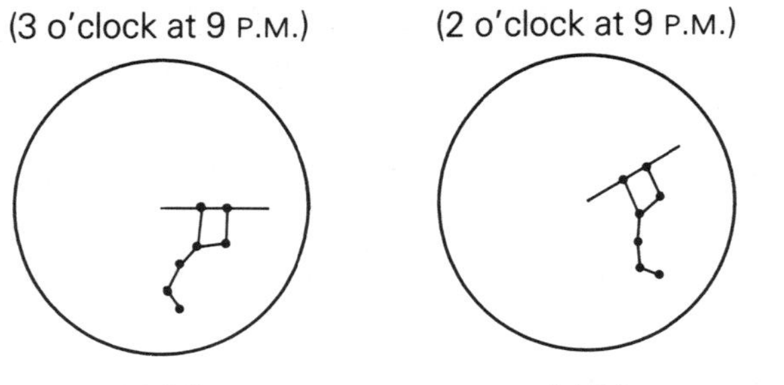

MARCH
(1 o'clock at 9 P.M.)

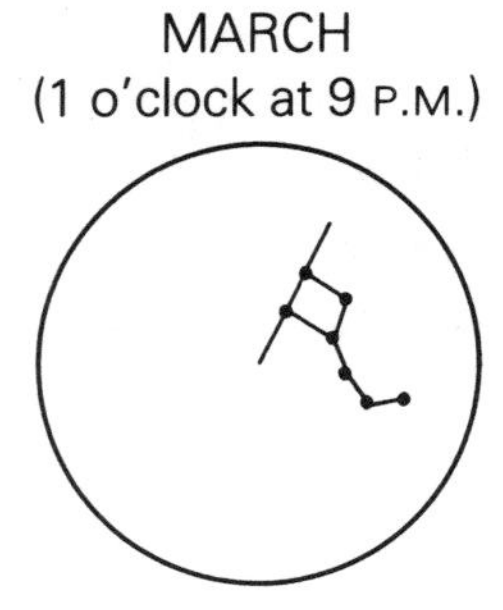

APRIL
(12 o'clock at 9 P.M.)

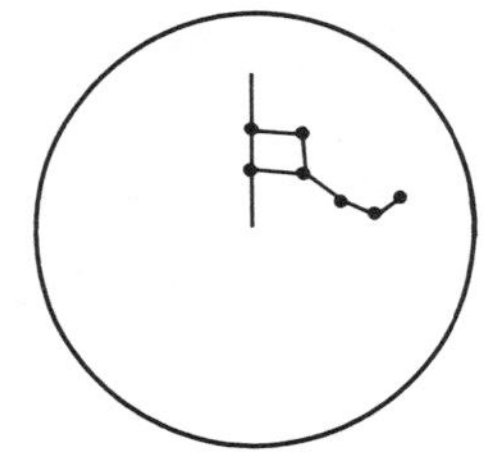

MAY
(11 o'clock at 9 P.M.)

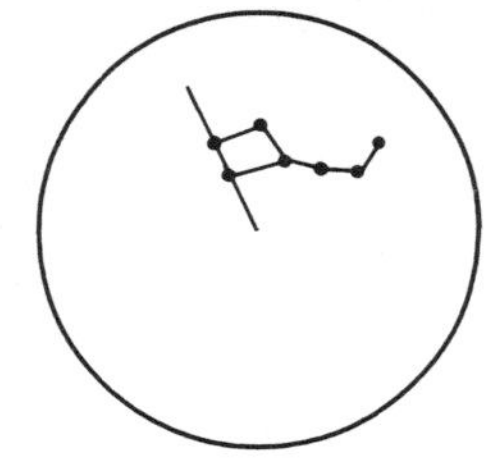

JUNE
(10 o'clock at 9 P.M.)

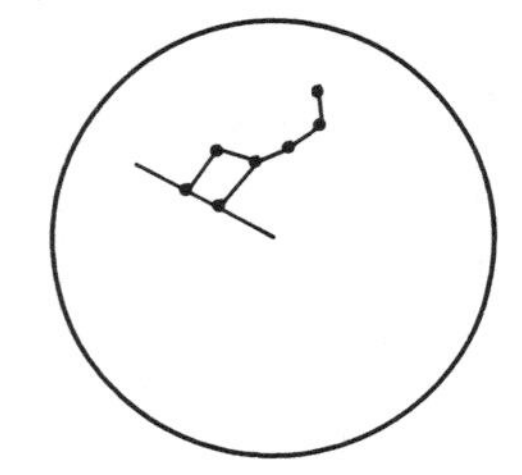

JULY
(9 o'clock at 9 P.M.)

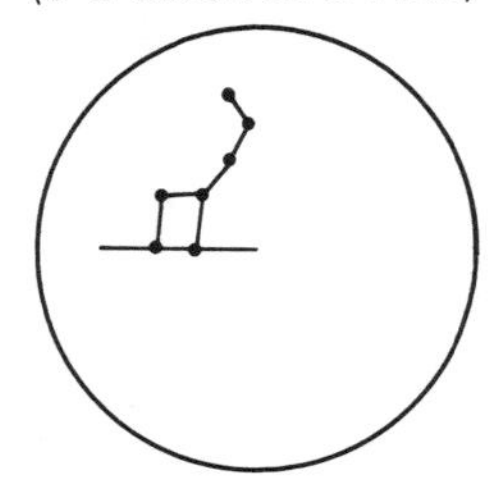

AUGUST
(8 o'clock at 9 P.M.)

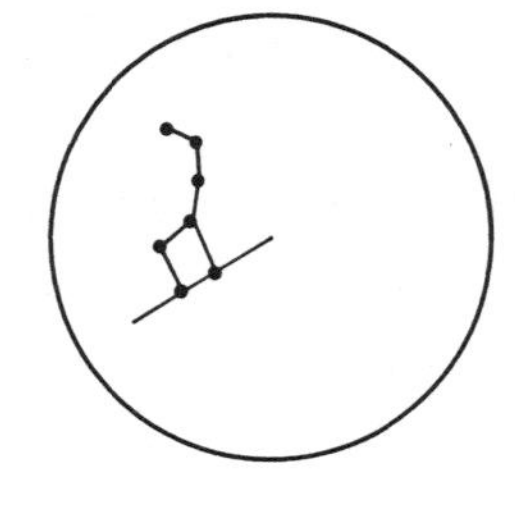

SEPTEMBER
(7 o'clock at 9 P.M.)

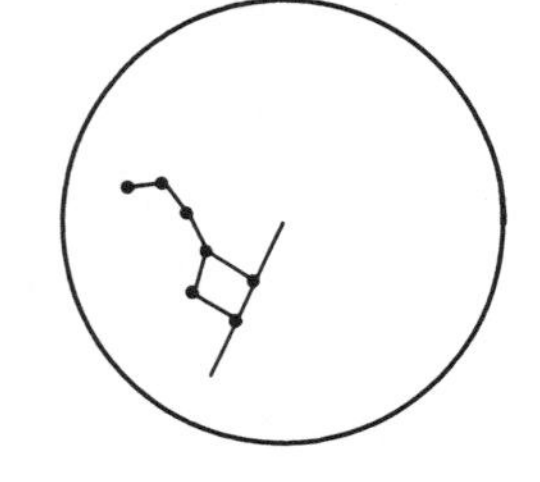

OCTOBER
(6 o'clock at 9 P.M.)

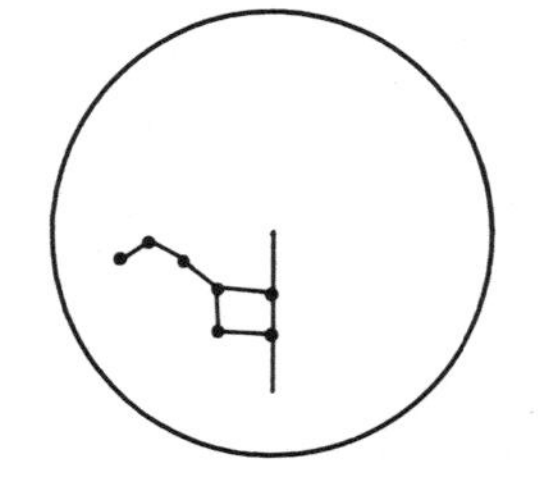

NOVEMBER
(5 o'clock at 9 P.M.)

DECEMBER
(4 o'clock at 9 P.M.)

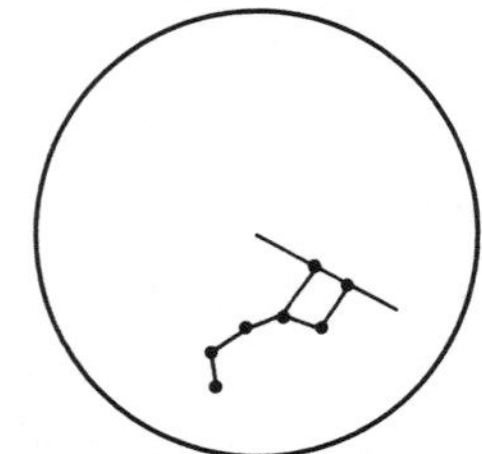

pointer stars of the Big Dipper on your sky clock at the 6 o'clock position, as shown in the next drawing. The inner numbered circle, which measures the 24 hours required for the earth to turn once on its axis, shows you that 6 hours (6 to 12) have passed since the pointer stars were at the 9 P.M. position, which is where they would be at 9 P.M. on July 15. Therefore, the time is 3 A.M.

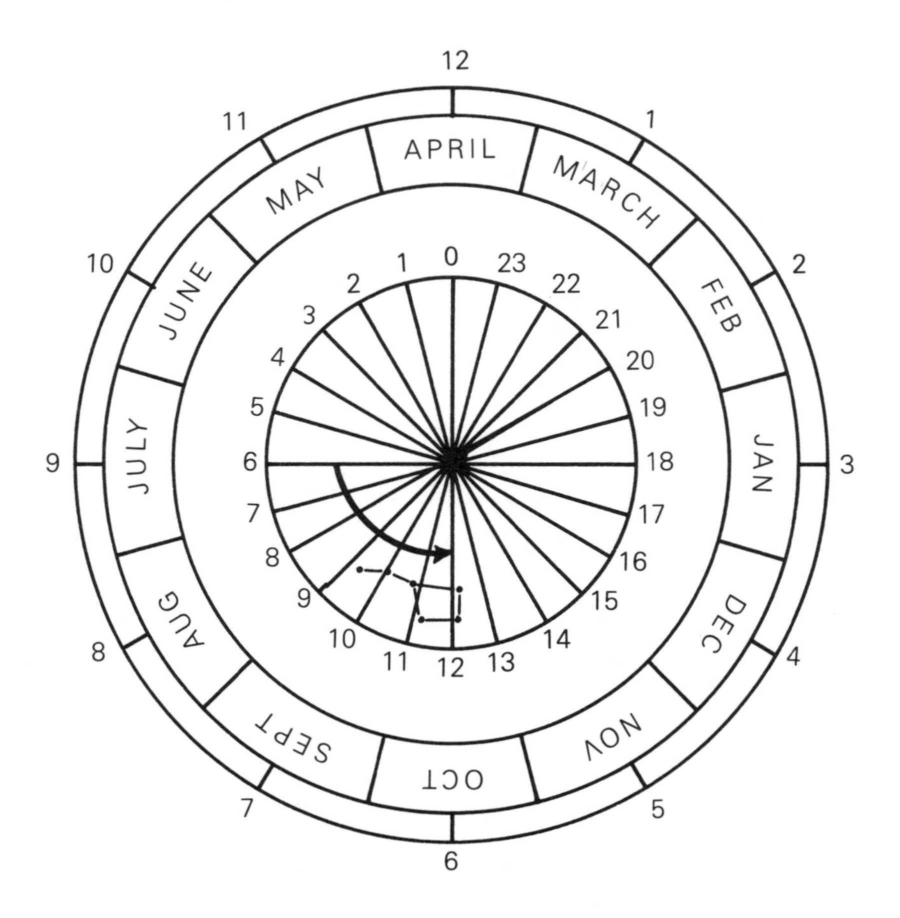


Now, suppose it's November 23, and you see the pointer stars in the same 6 o'clock position. On November 15, at 9 P.M., you'd expect the pointer stars to be in the 5 o'clock position. However, the inner circle shows you that these pointer stars are at a point 2 hours earlier than the 5 o'clock position (14 back to 12). If it were November 15, the time would be 7 P.M. Because it is about a week later than November 15, the clock position of the pointer stars at 9 P.M. will be at about 4:45. Consequently, the time is probably closer to 6:30.

With a little practice, you'll be able to estimate time quite accurately with your sky clock. If you live near the eastern or western end of a time zone, your clock will appear to be a little fast or slow, and you'll have to adjust your estimate accordingly. Since the sky clock is set for standard time, you'll have to add an hour if you are on daylight saving time.

## Polar Constellations

There are other bright constellations that turn about Polaris. The star chart shows you these constellations. The brighter the star, the bigger the dot on the chart.

Take the star chart outside. Bring a flashlight, but cover the glass in the flashlight with red cellophane. In that way, you won't lose your night vision when you shine the light on the chart. Place the appropriate month of the year uppermost against the northern sky to find the constellations. Can you find all the constellations on the chart?

## Other Constellations

The star charts on pages 40–43 show major constellations for the northern hemisphere at about the middle of the season indicated. Look in a direction — north, south, east, or west — as indicated by a star chart. The horizontal line at the bottom of each quadrant in the star chart represents the horizon. Depending on your latitude,

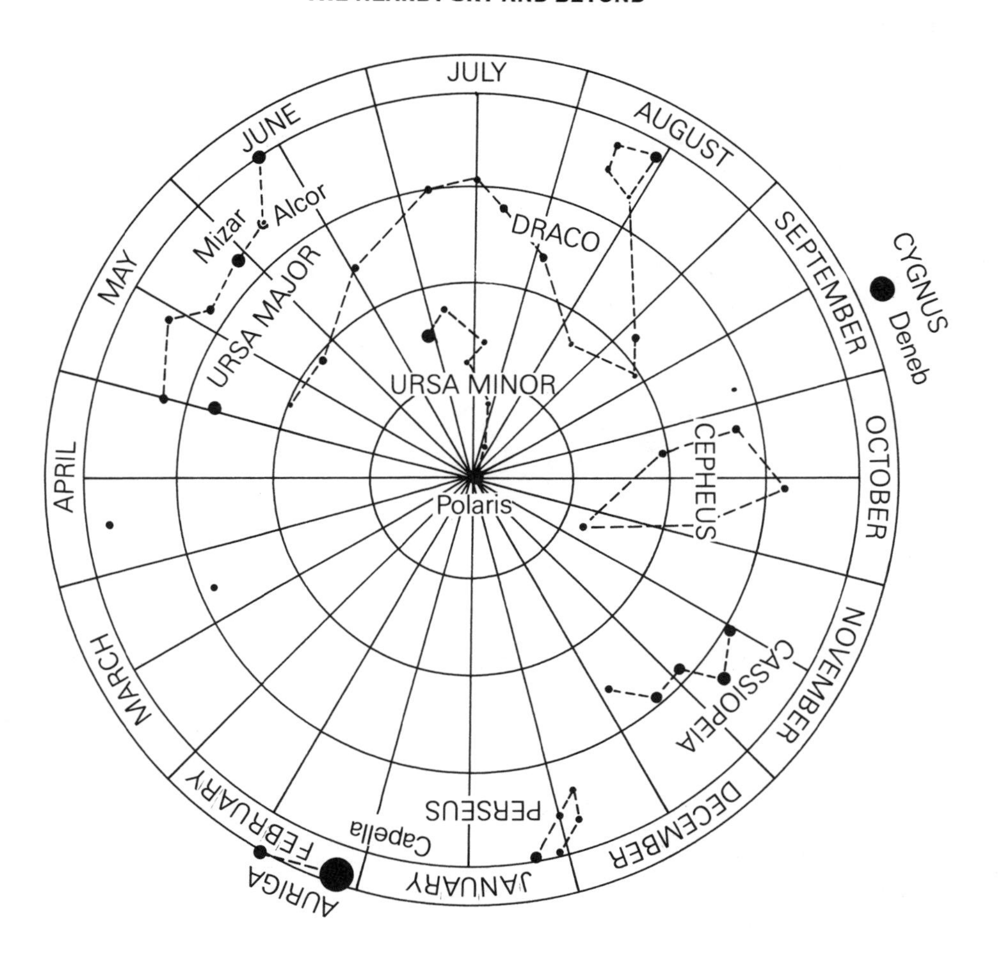


the view may be a little different than the one shown on the chart. There are star charts for the southern hemisphere too. If you travel below the equator, you will see stars that cannot be seen in the United States or Canada.

During the winter, Orion, the hunter, can be picked out quite easily. Betelgeux and Rigel are bright stars on either side of the three stars that make up Orion's belt. Canis Major, with the

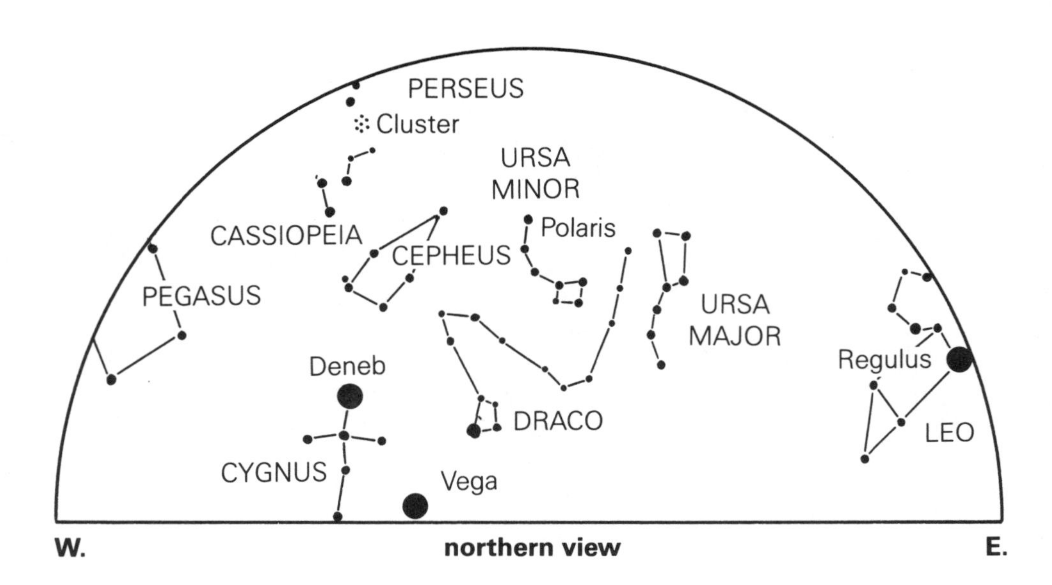


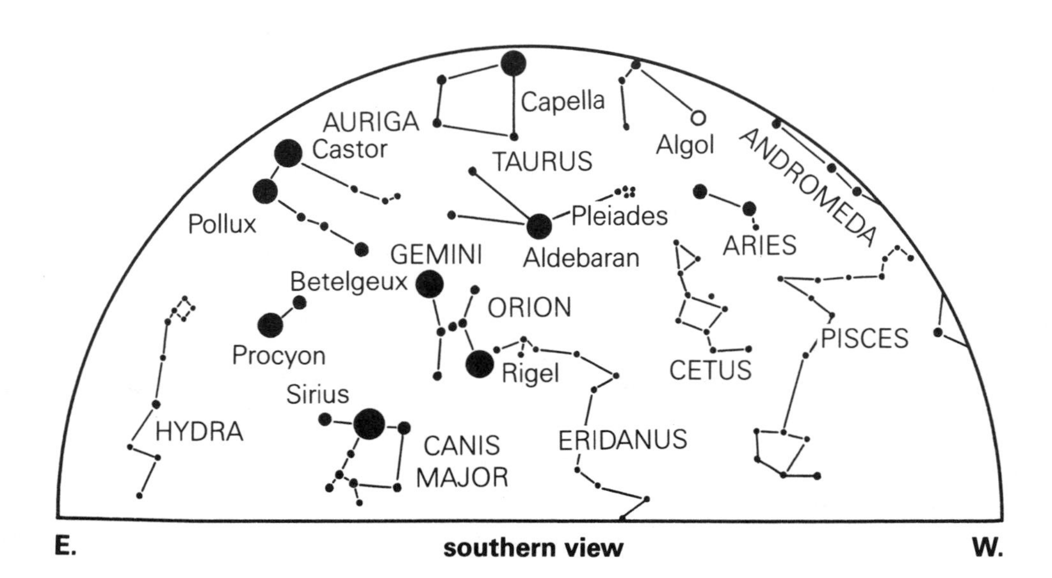


THE WINTER SKY

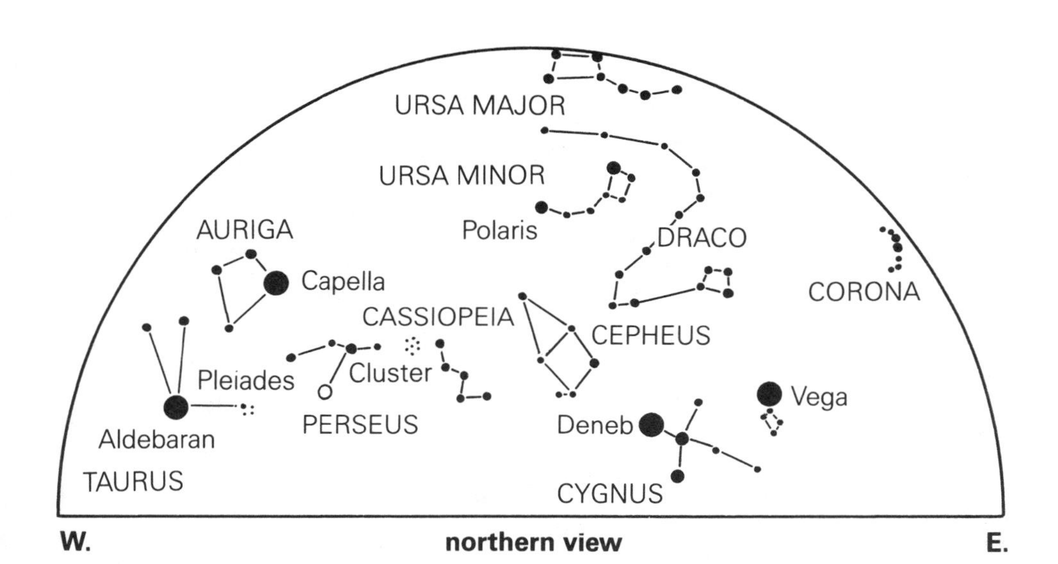


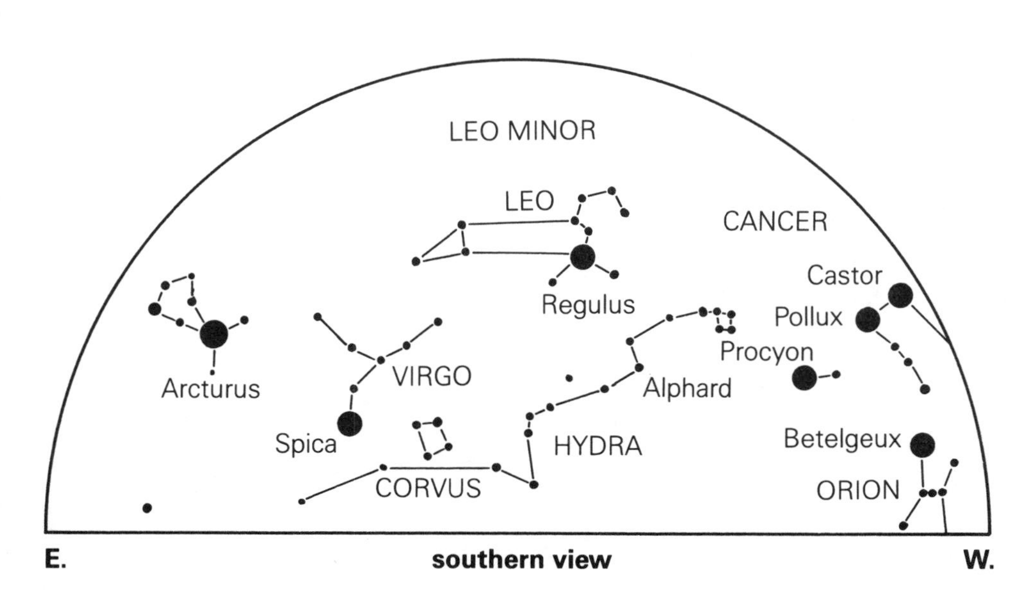


THE SPRING SKY

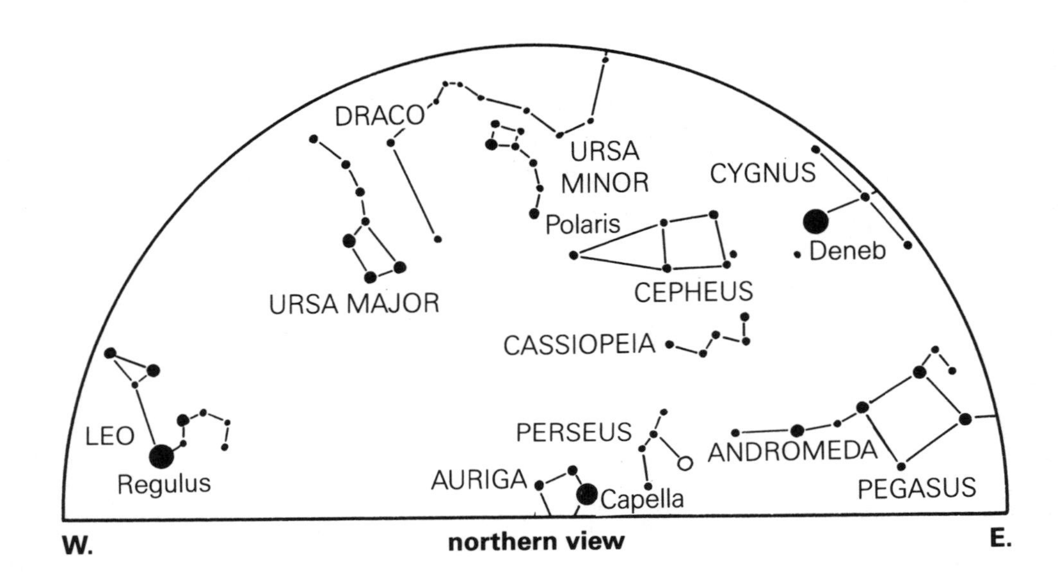


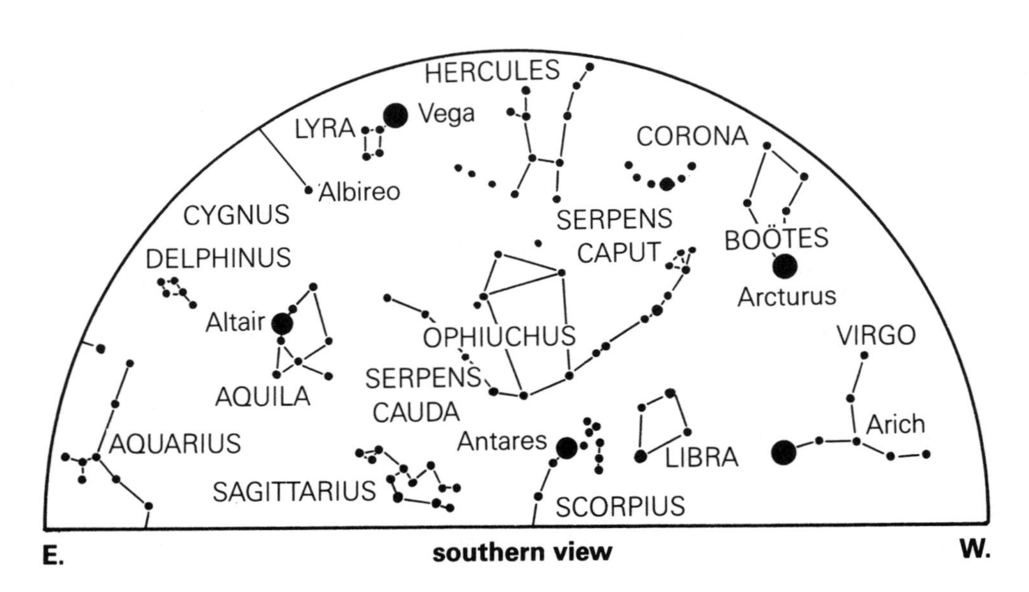


THE SUMMER SKY

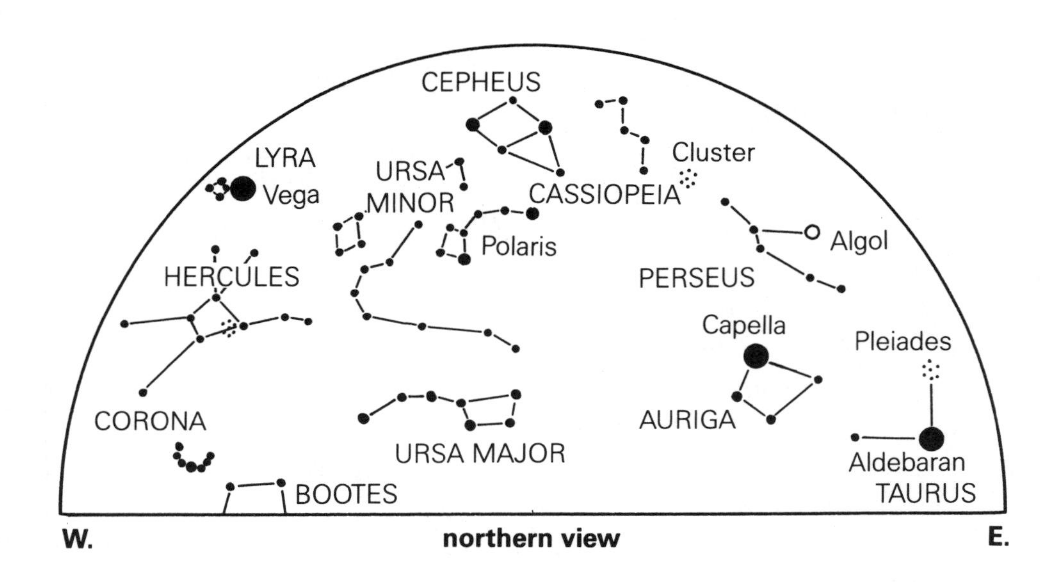


**northern view**

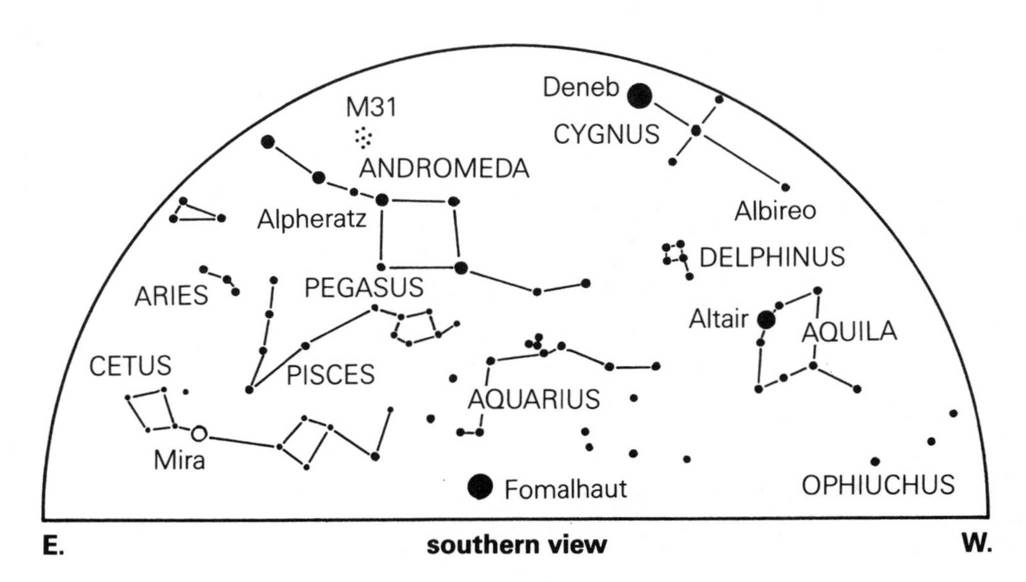


**southern view**

THE AUTUMN SKY

brightest star in the sky, Sirius, lies below Orion along the southern horizon. Capella, the bright star in Auriga; Gemini, with the twin stars Pollux and Castor; Taurus, with the bright orange star Aldebaran; and the tiny dipper-shaped group of stars known as the Pleiades lie overhead. Looking east to west, the square part of Pegasus is in the west as Leo rises in the east.

See if you can locate these constellations on clear nights in the winter sky.

At the beginning of spring, Orion may still be visible in the early evening but has moved southwestward as has Gemini. Auriga is in the southwest, and the bright blue star Vega, in Lyra, is in the east. Leo is high in the southern sky, and Boötes, a warped-kite shaped constellation with the bright, light orange star, Arcturus, can be seen in the southeast.

The summer sky is dominated by a few bright stars: Deneb in Cygnus, Vega in Lyra, Arcturus in Boötes, and Altair in Aquila. The coffee-pot-shaped constellation Sagittarius can be found on the southern horizon. West of Sagittarius lies Scorpius, with the bright red star Antares. West of Scorpius is Virgo, with its bright star, Spica.

As the Big Dipper sinks toward the northern horizon and Cassiopeia ascends towards its zenith on autumn evenings, the not so bright constellation Pegasus can be found high in the southern sky. Capella in Auriga can be seen in the northeast, and Vega in Lyra may still be seen in the northwest. Aldebaran and the Pleiades in Taurus may be seen rising along the eastern horizon.

If you enjoy stargazing and would like to be able to identify more constellations, you'll find more detailed star charts that you can buy or borrow useful in locating the various constellations. On the next four pages you will find some of the star charts from Levitt and Marshall's book *Star Maps for Beginners.* It has star charts for each month that are quite easy to read.

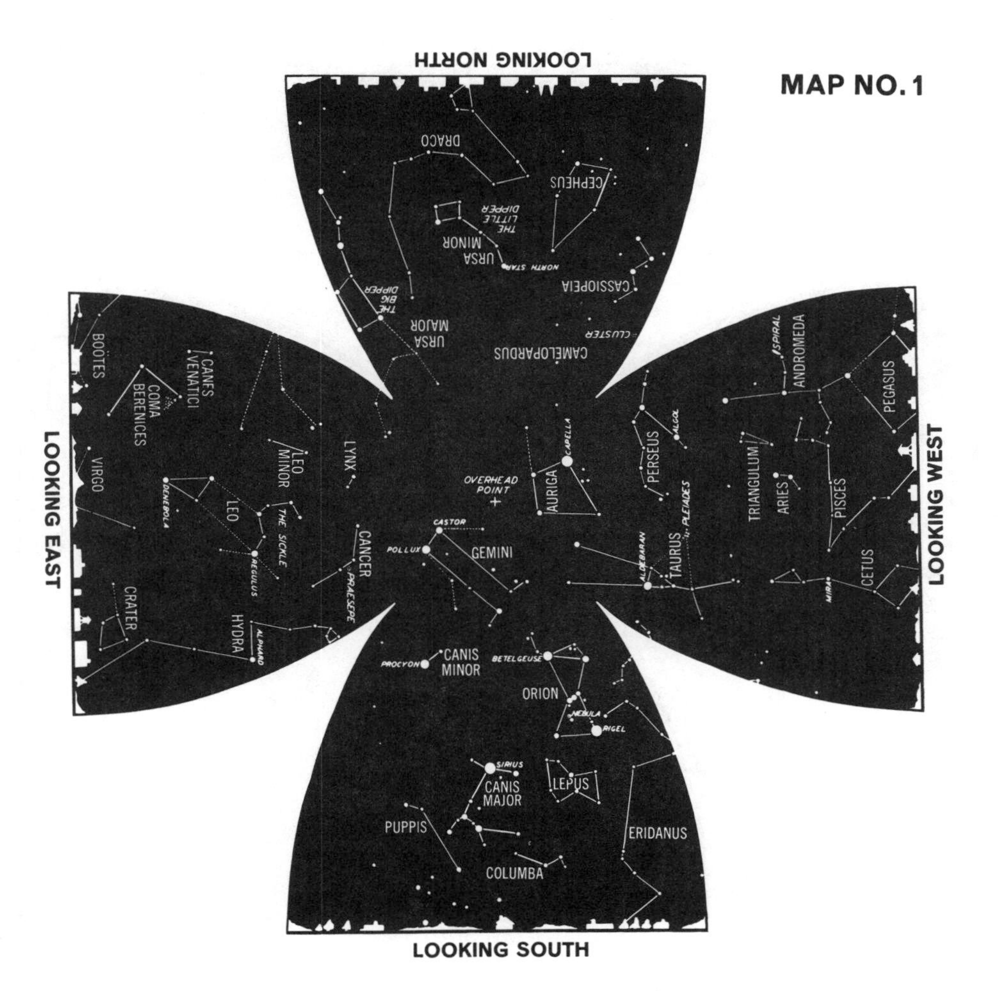


**This map represents the sky at the following standard times (for daylight saving time, add one hour):**

**FEBRUARY 1 at 10 p.m.**

**FEBRUARY 15 at 9 p.m.**

**MARCH 1 at 8 p.m.**

**This map represents the sky at the following standard times (for daylight saving time, add one hour):**

**MAY 1 at 10 p.m.**
**MAY 16 at 9 p.m.**
**JUNE 1 at 8 p.m.**

**This map represents the sky at the following standard times (for daylight saving time, add one hour):**

**AUGUST 1 at 10 p.m.**

**AUGUST 16 at 9 p.m.**

**SEPTEMBER 1 at 8 p.m.**

**This map represents the sky at the following standard times (for daylight saving time, add one hour):**

**NOVEMBER 1 at 10 p.m.**
**NOVEMBER 16 at 9 p.m.**
**DECEMBER 1 at 8 p.m.**

## Types of Stars

Stars are classifed according to their surface temperature. Astronomers can determine a star's temperature by the type of light it emits. Red stars are cooler than blue stars as you can see in Table 3.

**TABLE 3**

| Star Type | Temperature (°C) | Color | Example |
|---|---|---|---|
| O | >35,000 | greenish or bluish-white | rare |
| B | 10,000–35,000 | hot white | Rigel in Orion |
| A | about 10,000 | cool white | Sirius in Canis Major |
| F | 6,000–10,000 | slightly yellow | Polaris |
| G | about 6,000 | yellow | the sun |
| K | about 4,000 | orange | Arcturus in Boötes |
| M | about 3,000 | red | Betelgeux in Orion |

## A Spectroscope

To examine the light emitted by stars, astronomers use spectroscopes. A spectroscope is a device that separates light into the various colors that make it up. Light has wavelike properties, and we find that red light has a longer wavelength than blue light. To separate ordinary white light into its various colors or wavelengths, you can build a simple spectroscope as shown on the next page.

Cut two small square holes in opposite ends of a shoe box. Cover one of the holes with two small pieces of cardboard so that they create a narrow slit as shown. Tape a piece of diffraction grating over the other hole. But before you fix it in place, hold it up toward a

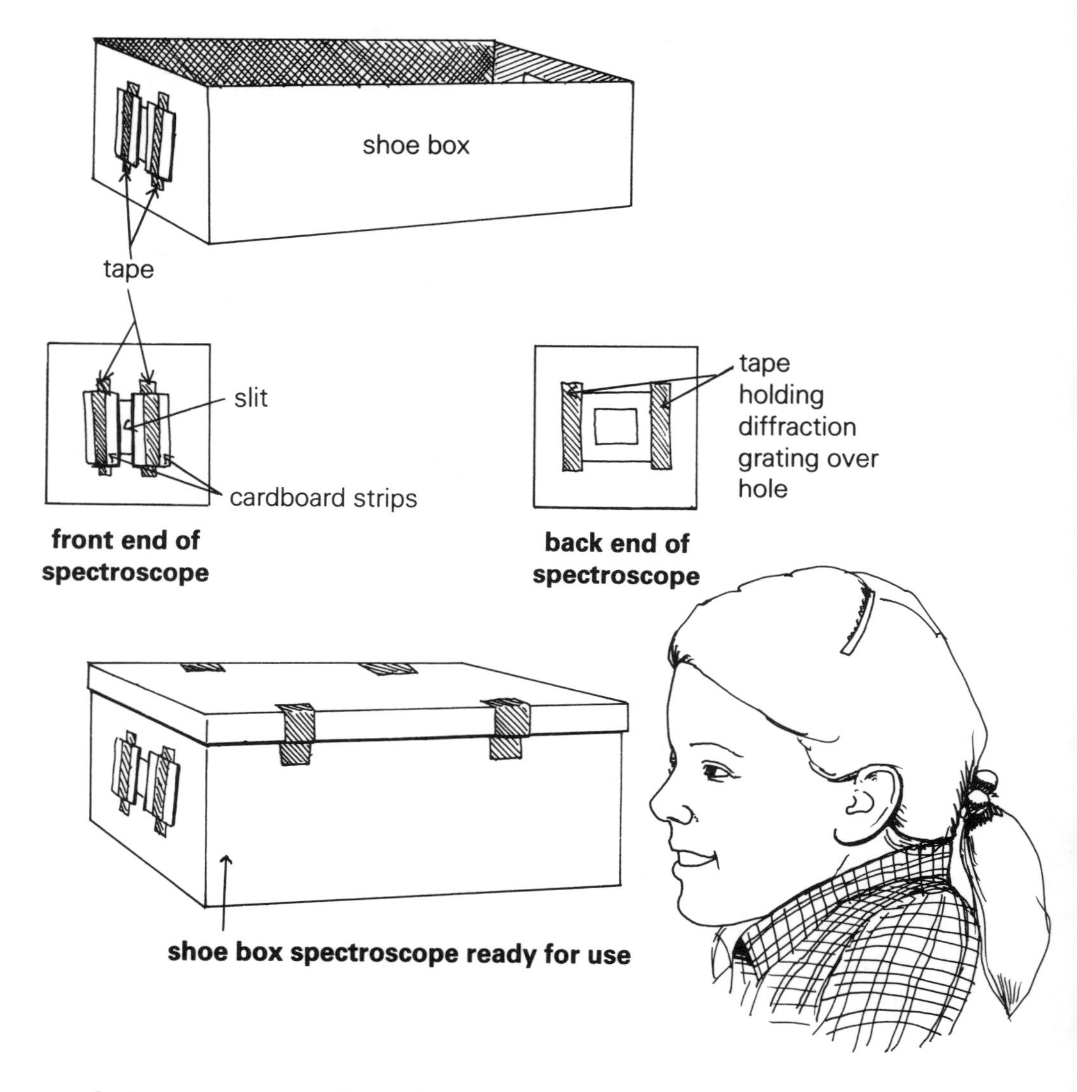


light. Turn it so that when light comes through the grating it spreads out into a *horizontal* spectrum of colors. You can buy diffraction gratings in a hobby shop or order them from one of the scientific supply companies found in the appendix. Cover the rest of the inside of the box with black construction paper and tape it shut.

Hold the spectroscope so that the slit in the box is parallel with the bright filament of a showcase lamp or an unfrosted light bulb. Look to either side of the slit: you will see a spectrum containing all the colors of the rainbow. Look at a fluorescent light bulb. You'll see not only a spectrum, but the bright violet, green, and yellow lines emitted by the mercury vapor inside the bulb. Like mercury, each element emits characteristic wavelengths or colors. Look at a neon light. What colors are released by neon? With spectroscopes, astronomers can figure out what elements are in stars.

To see the colors in sunlight, you'll need a cardboard box large enough for your upper body to fit in. Cut a small hole in the middle of one side, near the bottom. Tape a piece of diffraction grating over the hole. Then tape a sheet of white paper to the inside of the box on the side opposite the hole. Turn the box upside down and get inside with your back toward the diffraction grating. **Remember: Never look at the sun.** Have a partner help you turn the box so that sunlight falls onto the grating. Look on the white screen to see the colors found in sunlight.

To view a solar eclipse, substitute a pinhole for the diffraction grating.

When a star is coming toward us at high speed, the wavelengths of the light it emits are shorter than normal. If the star is traveling away from us, the waves of its light are "stretched" into longer wavelengths. This phenomenon is known as the Doppler effect. You've probably observed this effect with sound. If a car is approaching you with its horn blaring, the pitch seems higher than normal because the vibrations are more frequent than normal. The sound waves are "squeezed" together because the horn is traveling toward you as it emits sound waves. Thus, one wave follows the next sooner than it would if the car were at rest. When the car passes you and moves away, the pitch seems lower than normal. The car is moving away from the waves that strike your ear, so the waves are "stretched" and reach you less frequently than they normally would.

In the case of receding stars, astronomers say there is a *red shift* because the wavelengths of the light emitted are longer than they would be if the star were at rest. Similarly, approaching stars show a *blue shift*. Knowing the speed of light and the amount that known wavelengths emitted by various elements in the stars are shifted, astronomers can determine the speed that stars are moving toward or away from us. By comparing the red shifts or blue shifts of stars turning on their axes, we can tell which way a star is spinning. Further, from the red shifts seen in galaxies, there is good evidence that the universe is expanding and that the farther a galaxy is from us, the faster it is moving away.

## The Birth and Death of Stars

As gravity pulls together the dark thin globs of dust in space, the cold matter begins to get warmer. Eventually, temperatures rise to the point where hydrogen begins to fuse to make helium at the core of the contracting matter. At this point, the pressure created by energy produced at the core is enough to balance the inward pressure due to gravity, and the star becomes stable. If the contracting mass is less than 8 percent of our sun's mass, a solid, planetlike object will form. If the mass is greater than 100 suns, radiation emitted will prevent the contraction needed to ignite hydrogen and produce a star.

Big stars burn hydrogen faster than small stars. The lifetime of our sun is estimated to be about 10 billion years. A star with half the mass of the sun would live for 200 billion years. But one with a mass 15 times that of the sun would burn out in only 15 million years. The large stars are the O, B, and A stars in Table 3. Our sun is a type G star.

When all the hydrogen in a star's core has burned, hydrogen in the shell around the core begins to burn. Contraction, due to gravity, within the core makes the core hotter, causing layers around the star to expand. As the star expands, its surface temperature falls to about 4,000 degrees or less. Such stars, entering the last phases of their lives, emit red light and are called *red giants*. Betelgeux, in the constellation Orion, is an example of a red giant.

In another 5 billion years the sun will become a red giant, swelling until it swallows Mercury, Venus, and Earth and vaporizes the atmospheres of the outer planets.

The core of an old star in its red giant stage is rich in helium because helium is the product of hydrogen fusion. As the red giant's core contracts, its temperature rises still higher. When it reaches about 100 million degrees, helium begins to burn (fuse), producing heavier elements. A star's helium takes only about one-tenth as long to burn as hydrogen, so a star in the red giant stage is near the end of its life cycle. The final ashes of helium burning are carbon and oxygen. When the helium in the core has been consumed, a sequence similar to that following the burning of the hydrogen core takes place. The core contracts, and temperature increases from the contraction ignite the helium in the shell around the core and it begins to burn. The star expands again, producing a red supergiant. Pulses of energy from the helium-burning core cause as much as half of a star's mass to be blown off into space.

Ejection of the outer layers of a dying star expose its hot core. Ultraviolet light emitted by the core may cause the ejected gases to glow, producing a planetary nebula.

After puffing away much of its mass, the core of a low-mass star contracts to become a white dwarf. The matter in a white dwarf is packed together so tightly that it weighs about a million times as much as an equal volume of water. As the star cools over billions of years, it grows dimmer and finally becomes a dense, cold, earth-sized sphere.

Stars with masses greater than four times our sun's mass will contract again, raising temperatures to the point where carbon fuses, forming neon, magnesium, oxygen, and helium. Still more massive stars will fuse neon and helium into magnesium and fuse oxygen into sulfur, phosphorus, silicon, and magnesium. Stars that are twenty-five or more times as massive as the sun will fuse silicon into iron—the end product of nuclear fusion.

Contraction of the core of such a large star squeezes matter until it is more than 100 trillion times as dense as water. It is packed as tightly as the nuclei of atoms. As other contracting matter hits this dense core, it bounces off, producing a shock wave that blows the star apart.

## A Supernova

As the outer layers of a large star are blown apart, the star's brightness increases 100 million times in the course of seconds. The sudden appearance of such a bright star in the heavens is called a supernova. The appearance of a supernova in the southern hemisphere that was first seen on February 24, 1987, has excited astronomers throughout the world. The Supernova 1987A-Shelton, which is 170,000 ly away, was named after Ian Shelton of the University of Toronto. He was working at the Carnegie Institution of Washington's Las Campanas Observatory in Chile when he first saw the star. Though it burned with the brilliance of a billion suns and was clearly visible to the naked eye, it was not bright as stars go because it is so far away. However, SN 1987A is the first visible supernova since 1885 and the brightest since 1604. SN 1987A differs from other supernovas in many ways and is accompanied by an unexplained smaller companion object.

The core of a supernova consists of neutrons, because electrons in the compressed matter have combined with protons to produce neutrons. Physicists theorize that this core, which is a neutron star, should release neutrinos (tiny particles carrying little or no mass, and no charge) as the supernova forms. The fact that neutrinos were detected on Earth at about the same time that the supernova appeared seems to confirm the theory. It would also suggest that the neutrinos have no mass, since they traveled as fast as the light from the supernova.

Sometimes a star will suddenly increase a million times in brightness. Such a star is called a nova. Novas are more common than supernovas. Astronomers think they are formed when a white dwarf pulls matter from a larger companion star. As hydrogen is pulled inward by the smaller star, the temperature of the gas rises until it begins to undergo fusion, producing a gigantic hydrogen bomb.

## Black Holes

Calculations show that if the corpse of a dying star is more than three times greater than the mass of the sun, the escape velocity from this extremely dense body will be greater than the speed of light. This means that nothing, not even light, can escape from the dead star. Many astronomers believe such dense objects exist. They are called black holes. Black because no light can escape from them, and holes because anything coming close to their intense gravity will fall into them.

There is evidence to suggest that black holes do exist. Some stars seem to orbit something invisible, something that could have the same huge density as a black hole. Astronomers at Ohio State University have evidence that leads them to believe they have seen a star near the center of a galaxy being torn apart by a black hole.

Some astronomers suggest that black holes near the centers of galaxies may themselves coalesce into giant black holes that will "eat" anything that gets caught in their gravity.

## QUASARS

During the late 1950s and early 1960s, radio astronomers—astronomers who detect objects in the sky through radio waves rather than light—discovered a number of objects that had very strange spectra. The stars were called quasistellar radio sources because of their starlike appearance. This term was soon shortened to *quasars*. Analysis of the light and radio waves from these sources led astronomers to realize that the strange spectra were the result of huge red shifts. The red shifts indicated that quasars were farther away than anything that had ever been detected before—up to 18 billion light years. Further, quasars seemed to emit energies equal to a trillion suns; yet, they were no larger than our solar system.

To explain how an object as small as our solar system can emit energy equal to that of one hundred galaxies, some astronomers suggest that quasars are powered by supermassive black holes. These black holes are surrounded by vast masses of gases that are accelerated around the hole as they spiral inward before being swallowed.

## COSMOLOGY AND THE ORIGIN OF THE UNIVERSE

As astronomers study the stars and galaxies in the space that surrounds us, the red shifts of these luminous bodies make it clear that the universe is expanding. One model of this expanding universe can be made from ink spots on a balloon. As the balloon is inflated, the spots get larger and grow farther apart. The farther one spot is from another, the faster they separate. This corresponds to the greater red shifts that astronomers find in more distant galaxies.

No matter where you are located in this balloonlike universe, the rest of the universe is receding.

The model of the expanding universe is often referred to as the Big Bang theory. At time zero, matter was infinitely dense, and the laws of physics that we know today did not apply. In a very short period of time, the matter suddenly began to expand; hence, the Big Bang. After about a million years, expansion caused the temperature to drop to a point where matter, as we know it, could begin to form from the primitive particles.

If the universe is dense enough, gravity will eventually stop expansion, and the universe will contract back to its original state. Such a universe is said to be bounded or closed. Its contraction, or Big Crunch, will be followed by another Big Bang. If the density of the universe is equal to about 3 hydrogen atoms per cubic meter, then the universe will continue to expand, but at a rate approaching zero. If the density is less than this, the universe will continue to expand and is unbounded or open.

When astronomers add up all the mass in the matter they can see, there does not appear to be enough to prevent the universe from expanding forever. It appears to be unbounded. But astronomer Vera Rubin, in her study of the rotation rate of galaxies, has found that the rotation of the outer parts of some galaxies cannot be explained unless there is more mass than appears visible in the galaxies. Many astronomers believe that the extra mass exists as dark matter—gas, rocks, or black holes. Some believe it is in the form of neutrinos, although recent evidence indicates that neutrinos probably have no mass.

Though we may eventually know if the universe is open or closed, astronomers do not believe we will ever know the origin of the matter or energy that led to the Big Bang.

CHAPTER 3

# MAN INTO SPACE

On October 4, 1957, the Soviet Union sent the world's first manmade satellite, Sputnik I, into orbit. Most people were amazed to learn that Sputnik was moving about the earth at a speed of nearly 18,000 mph. Yet, Sir Isaac Newton had shown how satellites could be launched from Earth almost three hundred years earlier. All that was needed was the power to give the satellite the required speed.

After Sputnik, it was only a matter of time before man would journey into space. On April 12, 1961, the Soviet Union launched Vostok I. On board was Yuri Gagarin, the first man to orbit the earth. Within the same decade, on July 20, 1969, as millions watched on television, American astronaut Neil Armstrong stepped onto the moon and said, "That's one small step for man, one giant leap for mankind."

**American astronauts left their footprints in the moon's soil.**

Today, people orbit Earth frequently, and some spend extended periods in space laboratories. We talk of colonies in space where people may spend their entire lives, of mines on the moon, and of joint U.S.–Soviet journeys to Mars. In this chapter we will look at the science that led to space travel.

## Newton and Motion

Sir Isaac Newton was the first person to grasp the secrets of motion. And so it was he who first explained the pathway to space.

Through experimentation and careful creative thinking, Newton discovered three laws of motion. Before Newton, most people thought that an object moved only while a force (a push or a pull) acted on it. If the force was removed, the object would stop. Newton realized that a body stops not because there is no force on it, but because a force opposes its motion. Usually this force is friction—the force between two surfaces when they rub together.

In his first law of motion, Newton stated that a body in motion will maintain its motion in the same direction unless acted upon by an outside force. A moving object continues to move in a straight line at a constant speed unless it is pushed or pulled by some force.

If you roll a marble along the floor, it continues to roll long after you stop pushing it. Of course, it will come to rest after a while because there is some friction between the marble and the floor.

To see more dramatic evidence for Newton's first law of motion, you can build a frictionless car.

## A Frictionless Air Car

To build a small frictionless "car" you will need a balloon, a wooden spool (the kind thread comes on), a piece of 1/4-inch-thick plywood or Masonite, glue, and sandpaper. **Ask an adult to help you** cut a square about 2½ inches on a side from the wood and drill a 1/16-inch hole through the center of the square.

Make one side of the wood very smooth by rubbing it with sandpaper. The lower edges and corners of the wooden square should be rounded and smoothed with sandpaper too. Fasten the spool to the unsanded side of the square with glue. The holes in the spool and wooden square should line up.

After the glue has dried, an inflated balloon can be attached to the spool. Air will pass out of the balloon and through the holes in the spool and wood, forming a thin layer of air beneath the wooden square. This thin air layer reduces the friction between the wood and the surface across which it moves.

Place the air car on a smooth level surface such as a formica-covered counter. Give it a small push and watch it go. Does it show any sign of slowing down? What happens to the car when the balloon is empty? How does this moving air car illustrate Newton's first law of motion?

If you have access to an air hockey table, you can experiment there too. Why do the pucks move at constant speed across the table? What happens to them when the air pump is turned off?

Seat belts are a practical application of the first law of motion. To see why, place a doll on a large toy truck. Give the doll and truck a push along a level floor. Arrange for the truck to slam into a pile of bricks. What happens to the doll when the truck collides with the bricks? Now, attach the doll to the truck with a strong rubber band. The rubber band represents a seat belt. Repeat the experiment. What happens to the doll this time? How are seat belts related to the first law of motion? How do they save lives?

## MORE ON THE FIRST LAW OF MOTION

There is a second part to Newton's first law. Suppose an object is at rest—has a speed of zero. Newton's law says it will retain that speed unless acted upon by an outside force. In other words, a body maintains its state of motion, rest, or velocity unless a force is applied to it. Things tend to stay the way they are. They resist any attempt to change their present state. That's why the first law of motion is often referred to as the law of inertia (resistance to change).

A magician pulling a tablecloth from beneath dishes illustrates inertia. If frictional forces between cloth and dishes are small, and the cloth is pulled quickly so that the forces act for a very short time, the dishes will remain in place. You can try the old tablecloth and dishes trick for yourself, but **use plastic dishes and protective goggles** until you become an expert. Place a smooth cloth or several sheets of newspaper on a small table. Set the table, in the kitchen, basement, or out of doors, with plastic ware, including a cup of water. Give the cover a quick yank and the dishes will remain in place. Be sure to pull the cover straight out, don't yank it up or down.

To do a simpler version of the tablecloth trick, cut an index card in half and center it over the mouth of a soda bottle. Place a marble on the card so that it is above the opening in the bottle. Give the card a sharp blow with a snap of your fingers. What happens to the card? What happens to the marble?

## When Forces Act

As you have seen, an object left alone does not change its state of motion. It will either remain at rest or keep on moving unless given a push or a pull. But what happens when a force acts? To find out, place the air car you made earlier on a smooth ramp. If you have a smooth table, you can make a ramp by placing blocks under the legs to raise one end of the table. Inflate the balloon and place the air car at the top of the ramp. When you release the car, the force of gravity acting along the ramp will be unopposed.

What happens to the car when you release it? Does its speed change? An increase in speed is called acceleration. Does the air car accelerate?

If you drop a ball, gravity will pull it straight toward the center of the earth. Do you think the ball will accelerate? If it falls at constant speed, how will the time to fall one yard compare with the time to fall two yards? To fall four yards?

To measure the time the ball falls, count to five as fast as you can. You'll find it takes just about one second to say, "One, two, three, four, five," as fast as you can. If you get to three when the ball hits the floor, it fell for three-fifths of a second. If you count to five and then get to two on your second count, the ball fell for one and two-fifths seconds.

Drop a ball several times from heights of 1, 2, and 4 yards. Measure the time by the count method for each trial. Does a ball fall at a steady speed under the force of gravity? Or does it accelerate as the air car did?

What happens to a body when a force other than gravity acts on it? Will it accelerate as the air car and ball did? To find out, place pieces of tape at different distances from a starting position along a long, level hallway or gymnasium. The starting position should be marked 0. Then place markers at 5 yards, 10 yards, 15 yards, and at farther points if possible.

You know that if an object moves at constant speed, it will take twice as long to go 20 yards as to go 10 yards. On the other hand, if an object accelerates, it takes less than twice as long to go twice as far.

To see what happens when a force unrelated to gravity is applied to a body, have someone sit in a wagon or on a skatebord. He or she holds one end of a spring scale firmly, as shown in the drawing on the next page. Find the force that is needed just to keep the wagon or skateboard moving at a very slow speed. This is the force needed to overcome friction between the wheels and the floor.

Place the vehicle at the 0 position of your marked hallway. Have another person hold the skateboard or wagon. The person on the wagon or skateboard should hold the spring scale firmly as before. Meanwhile, you pull on the other end of the spring scale until you

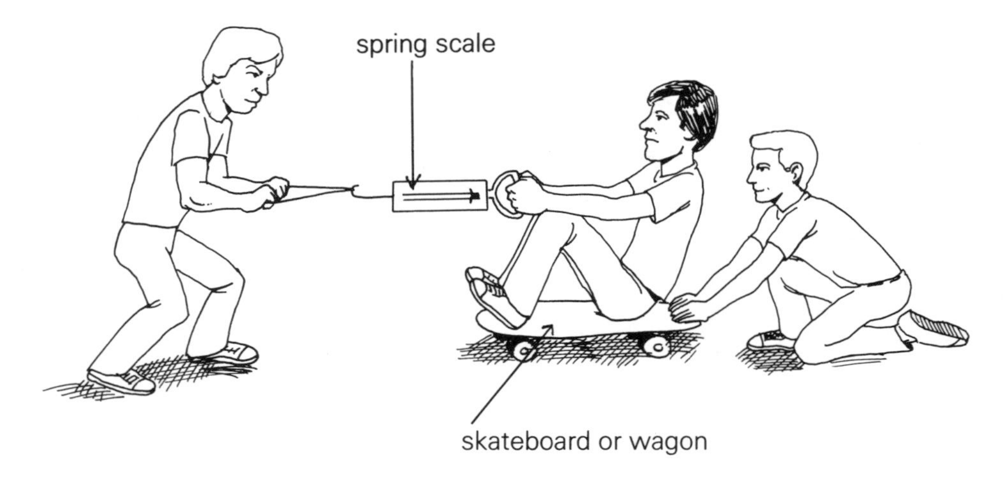


are applying a force greater than friction. The person holding the wagon or skateboard should now let go. You must keep the force constant as you pull the wagon along the hallway after it or the skateboard is released. You may have to experiment a few times to find a force that you can keep constant all the way along the measured path.

Measure the time it takes to go the various distances. Does the wagon accelerate when a force is applied?

What happens to the acceleration if you make the force larger?

What happens to the acceleration if the same force is applied to someone much heavier sitting on the wagon or skateboard? What happens if a much smaller person is pulled along the path with the same force?

## Newton and the Second Law of Motion

Newton found that when any kind of force acts on an object, the object accelerates in the direction of the force. Of course, the force has to be greater than any force opposing the motion such as friction. If a force of 10 pounds is applied to an object and there is a frictional force of 4 pounds, only 6 pounds of force, the net force, will cause acceleration. The other 4 pounds is required to overcome friction. In your experiment, if you applied 4 pounds of force just to keep the wagon moving at a constant speed, then the frictional force between wheels and floor was 4 pounds.

Newton found that if he doubled the net force acting on a body, the acceleration doubled too. He discovered too that if he doubled the mass, he had to double the net force to have the same acceleration. If he kept the net force constant and doubled the mass, the acceleration was halved. To summarize his results, we can say that the net force applied to a body is proportional to the mass times the acceleration: $F = ma$ (force = mass x acceleration). Or $a = F/m$ (acceleration = force/mass).

Since gravity acts toward the center of the earth, objects accelerate downward when released above the ground.

## GRAVITY AND MASS

Find two balls that have different masses. You might use a tennis ball and a baseball, or large and small lumps of clay. The ball that feels heavier has more mass.

Newton reasoned that the gravitational force the earth exerted on bodies near its surface was proportional to their mass, the amount of matter in the body. If body A has twice the mass of body B, the force of gravity on A will be twice that on B. If the mass of a body doubles and the force doubles too, the acceleration should remain the same since $a = F/m$.

To test Newton's idea, hold the two balls or lumps of clay at the same height above the floor. Release them together. Do they both strike the floor at the same time? Do they have the same acceleration?

To test the acceleration of two very different masses, hold a book several feet above the floor. The book's cover should be parallel to the floor. Place a sheet of paper that is smaller than the book on its cover. Release the book from both hands at the same time. Does the paper fall with the same acceleration as the book?

If you hold the book in one hand and the paper in the other, do they fall with the same acceleration when you release them? They would if they were dropped in a vacuum, but in air, the resistance of the air retards the paper more than the book. Wad the sheet of paper into a small ball and drop it beside the book. Do they fall together now?

Try a table tennis ball and a golf ball. **Ask an adult to help you** by simultaneously dropping a golf ball and a ping pong ball from a second story window. You can watch from the ground. Which ball has the greater acceleration? Why?

On one Apollo mission, an astronaut dropped a hammer and a feather at the same time. A camera recorded their side by side motion as they fell to the lunar surface. Though they fell with the same acceleration, the acceleration was only one-sixth the acceleration of falling bodies on Earth. The moon's gravity is weaker than Earth's because it has less mass.

Newton was able to show that the force of gravity between two masses is proportional to the masses of both bodies. Double the mass of either, and the force of attraction doubles. Double the mass of both, and the force will be four times as large. But if the distance between the centers of the masses doubles, the force is only one-quarter as large. If the distance between the masses is halved, the force quadruples.

The earth's mass is about 82 times that of the moon. So on the score of mass, our gravity should be 82 times as big as the moon's. However, the radius of the earth is about 3.7 times as large as the moon's and the force of gravity varies inversely with the square of the distance. So on the score of distance, the earth's gravity, if it had the same mass as the moon, would be 1/14th as big as the moon's ($\frac{1}{3.7}$ x $\frac{1}{3.7}$). When you multiply 82 by 1/14, the product is quite close to 6.

## NEWTON AND THE THIRD LAW OF MOTION

Put a rubber band around your two index fingers. Now stretch the rubber band. How do the forces on each finger compare?

Have a friend hold one spring scale while you hold an identical one that is attached to the one your friend holds. Tell your friend to hold his or her spring scale still while you pull on the other one. How does the force that you exert on your friend compare with the force that your friend exerts on you?

If you can find two skateboards, have a friend sit on one while you sit on the other. Both skateboards should be on a smooth level surface in a hallway, gymnasium, or large room. (If you can't find skateboards, this experiment can be done on roller skates, or on ice skates in a hockey rink.) Sit behind your friend with both skateboards pointed in the same direction. Give your friend a fairly good push with your hands. Does your friend move? Which way does he or she go? Did you move? Which way did you go?

The same experiment can be done on a smaller scale using a pair of toy trucks and a clothespin as shown in the drawing. Tape the

clothespin to one of the trucks. Push the trucks together, extending the spring in the clothespin. What happens when you release the trucks and the spring contracts? If you add mass to one truck, how does that affect the results of the experiment?

The experiments you have just done illustrate Newton's third law of motion: when one body exerts a force on a second, the second exerts an equal but opposite force on the first.

You pushed on your friend who accelerated away from you. Your friend, in turn, without even trying, exerted an equal but opposite force on you. Consequently, you accelerated in the opposite direction.

If you repeat the experiment with someone who has much less mass than you do, what do you think will happen? What will happen if your partner is more massive than you?

Do the experiment again, but this time push against a wall. It's clear there's a force on you because you accelerate away from the wall. But where is the opposite force—the one you exerted on the wall? The wall is attached to the earth; when you push on it, you are pushing on the earth. Because the earth's mass is huge compared to yours, its acceleration is too small to be observed.

The same is true when you release a ball that falls to the ground. The earth pulls on the ball and it accelerates downward. But the ball pulls the earth upward. However, the mass of the earth is so large that its acceleration can't be detected.

## Motion and Momentum

All motion involves Newton's third law. When you walk, you push backward against the ground. The ground pushes back on you but in the opposite direction. Therefore, you move forward. The earth, in turn, moves backward, but its mass is so large that we don't see its tiny movement.

**Ask an adult to help you** set up this experiment. It will convince you that you do push the earth opposite to the direction you walk. Place a long, wide, flat board on a number of rollers, such as wooden dowels, that rest on a level floor. **Have two adults, one on each side of you, hold your arms as you try to walk along the board.** You'll find the board moves backward as you try to walk forward. If you've ever tried to step onto a dock from a boat that wasn't tied to the dock, you've had a similar experience. As you stepped toward the dock, the boat moved away from the dock, leaving you, perhaps, in the water.

As you walked along the board, pushing backward on it, it pushed forward on you. You moved forward; it moved backward. Both you and the board acquired *momentum*. The momentum of a body is defined as its mass times its velocity. You acquire momentum because you accelerate to a certain speed while a force is applied to you. When you walked on the board, the board pushed you forward as you pushed it backward. As long as you push on it, it pushes back with an equal and opposite force. Therefore, both you and the board acquire equal but opposite momenta. If you and the board have the same mass, you will move in opposite directions, but with the same speed. If the board is half as massive as you, it will move twice as fast as you, so that your momentum will still be equal but opposite to the board's. Therefore, the sum of your and the board's momenta will be zero ($2m \times v/2 + (-mv) = 0$.)

Momentum is always conserved; that is, the total momentum never changes. For you to acquire momentum in one direction, you have to give something else an equal momentum in the opposite direction. You do that by exerting a force on that something else for some time. It exerts an equal but opposite force on you for the same time.

## Projectiles and Satellites

If you lift a ball and release it, it accelerates toward the center of the earth because of the force of gravity. But suppose you throw the ball horizontally, will it still accelerate toward the ground in the same way? To find out, place a coin near the edge of a table. Place a second coin on the end of a ruler as shown in the drawing. If you hit the ruler sharply at point F while holding the center of the ruler at point M with your finger, the ruler will swing about your finger.

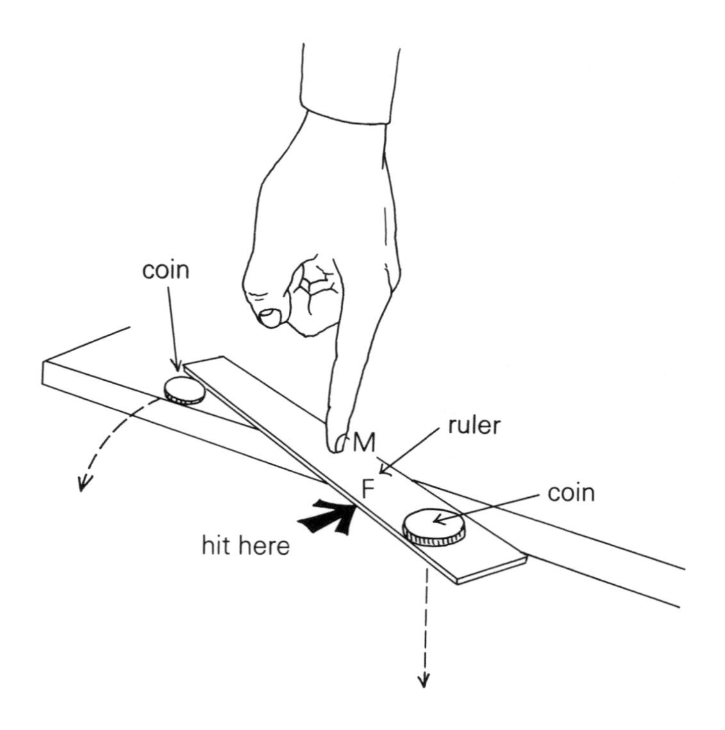


The coin at the table's edge will fly off in a horizontal direction. At the same time, the coin on the end of the ruler will fall straight to the floor. Listen carefully! Do you hear one sound or two when the coins strike the floor? What does this tell you about the downward acceleration of the two coins? About the downward acceleration of a ball that is thrown horizontally?

You've seen that an object's acceleration toward the earth is the same whether it falls straight down or travels horizontally as it falls. To see how its speed affects the path of a ball launched horizontally, build a ramp like the one shown in the drawing. Tape a long sheet of

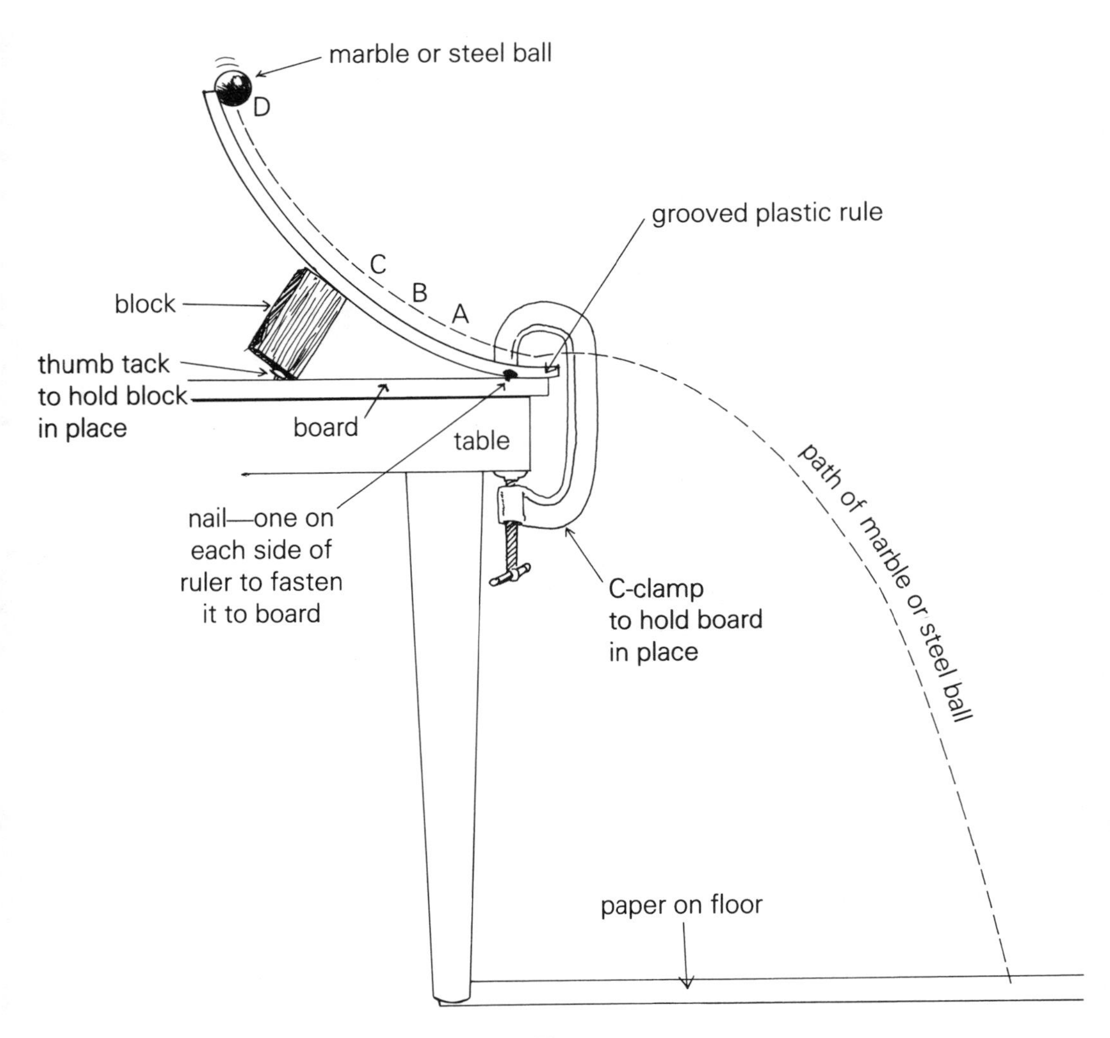

paper on the floor beneath the ramp. Mark the point where a marble lands if it falls from the end of the ramp. To give the marble some horizontal speed when it reaches the end of the ruler, release it from various points (A, B, C, D) along the ramp. As you see, the higher the point from which it is released, the faster will be its horizontal speed when it leaves the ruler. Mark the landing points of the marble as its horizontal speed increases. How does its horizontal speed affect its flight path?

Newton realized that if an object could be launched horizontally from a very high point above the earth, where air resistance was negligible, its path would become longer and longer as its speed was increased. Finally, at a speed of about 18,000 mph, the object's rate of fall would match the curvature of the earth as shown. The earth's surface curves 16 feet (4.9m) for every 5 miles (8 km) of horizontal distance. Thus, since an object falls off its horizontal path 16 feet in one second, its path will match the earth's curvature if it is traveling with a horizontal speed of 5 miles per second (18,000 mph or 29,000 kph).

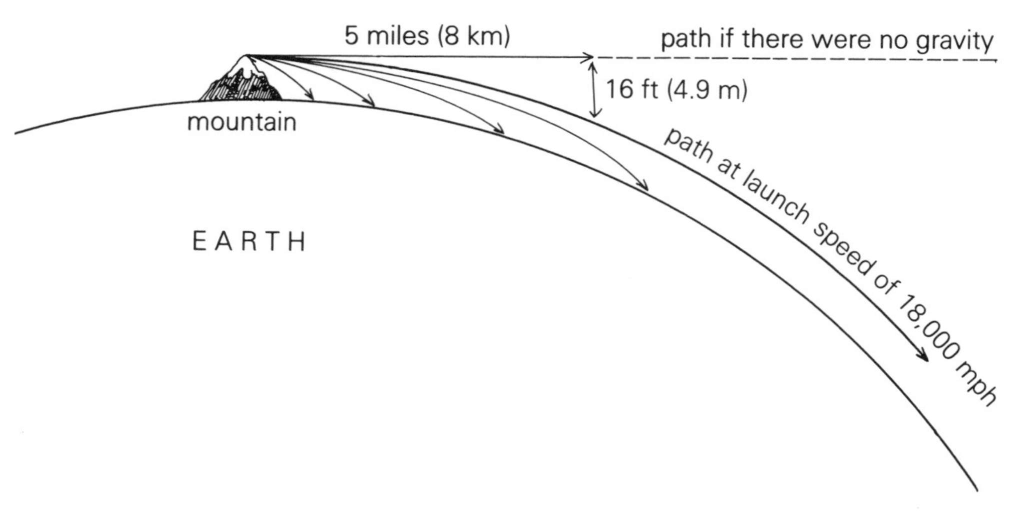

Newton knew how to put satellites into orbit about the earth. Sputnik would have been no surprise to him. What Newton lacked was the technology to accelerate objects to speeds of 18,000 mph.

## Rockets and Satellites

Rockets that send satellites into orbit use the principle of momentum to develop the necessary speed of 18,000 mph. By pushing fuel out the back of the rocket at high speed, the rocket acquires an equal momentum in the opposite direction.

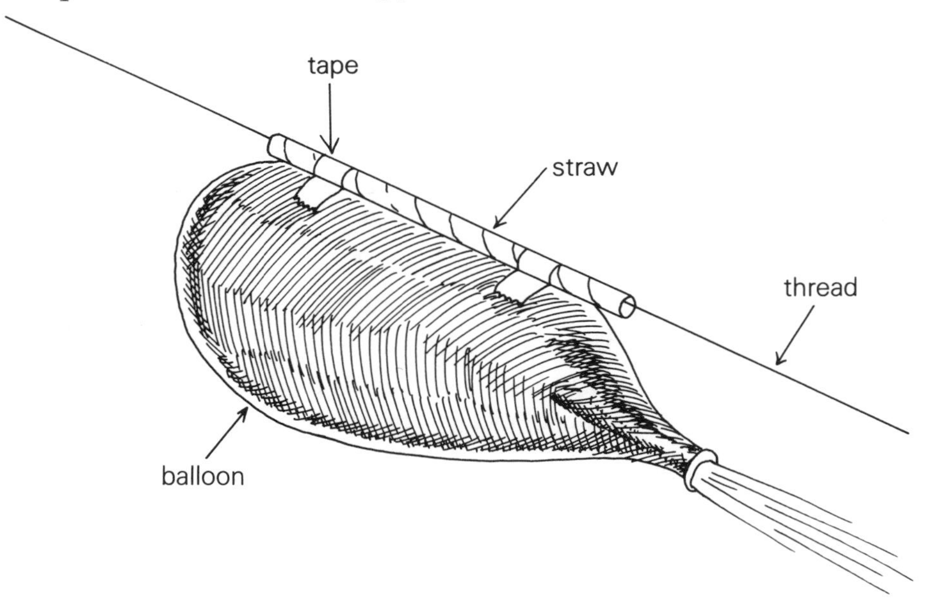


You can make a simple rocket as shown in the drawing. Tape a soda straw to a balloon, preferably an oblong one, and mount the soda straw on a long thread. When you release the balloon, the air pushed out of the balloon's neck will provide an equal force on the balloon. The momentum to the rear acquired by the air will equal the forward momentum supplied to the balloon rocket.

In principle, all rockets operate in this manner. However, the mass and speed of a satellite is much greater than the momentum of the balloon you launched. Fuels that can provide lots of momentum must be used to send satellites into orbit.

Little was done to develop the technology needed to launch the satellites that Newton knew were possible until Robert H. Goddard began building rockets in the 1920s. Goddard used gasoline to fuel his first rockets, but he realized that once rockets left the earth's atmosphere and entered space, they would have to carry their own oxygen to burn the fuel.

People laughed at Goddard because they heard he was building a spaceship to go to the moon. But Charles Lindbergh reconized the significance of Goddard's research and helped him find financial support. During the thirties, Goddard patented the idea of multistage rockets. He realized that it was foolish to lift the entire rocket into space. So he designed rockets in stages. Once the fuel in each stage had been used, that part of the rocket would fall back to the earth. By using stages, less fuel is needed because the mass being lifted is reduced after each stage. This principle of multistaging is still used in rockets today. The space shuttle discards its solid-fuel rocket boosters and its external fuel tank as it ascends to orbit.

As Goddard anticipated, the space shuttle carries its own supply of oxygen into space. The oxygen is in a liquid state under very high pressure. It is used to combine with liquid hydrogen. The reason for choosing hydrogen as the fuel is simple. Hydrogen is the lightest gas known, and it produces the most energy per mass of any fuel. By burning hydrogen, the space shuttle lifts as small a mass of fuel as possible. And the burned fuel leaves the rocket engine with the greatest momentum possible, providing the shuttle with an equal momentum in the opposite direction.

# CHAPTER 4

# ROUND AND ROUND WE GO

Since the weather satellite Tiros I was launched on April 1, 1960, meteorologists have been able to base their predictions on weather systems that they can see on a global scale. Most television weather reports include satellite photographs of the clouds and weather systems over the entire country. In addition to visible cloud cover, weather satellites provide meteorologists with information about soil moisture, atmospheric temperatures, ozone levels, gas and aerosol concentrations, and rainfall over the oceans.

Other satellites take photographs that enable scientists to prepare maps of unexplored areas, analyze forest, mineral, soil, and water resources, and make crop and earthquake forecasts. Communication satellites make it possible to send TV, radio, and telephone signals from one continent to another. Since 1964, we

An Intelsat Satellite prior to launch. It was placed in synchronous orbit where it allows TV signals to be transmitted between continents.

have been able to transmit TV broadcasts of Olympic Games and other events of worldwide interest across the globe so they can be viewed live.

## THE EARTH IS A SPHERE

People have known for centuries that the earth is round, but the spectacular views from satellites, and especially from the moon, should convince even members of the Flat Earth Society that the earth is a sphere.

**Earth as seen from the Apollo 17 spacecraft.**

Early evidence about the shape of the earth was more subtle. For example, during an eclipse of the moon, the earth's shadow as it crosses the moon was seen to be curved. As early as the third century B.C., Eratosthenes estimated the radius of the earth. He knew that on the first day of summer, the sun was directly overhead in the city of Syene (now Aswân). He knew this because the image of the sun could be seen reflected from the water in a deep well in Syene on that day. Eratosthenes lived in Alexandria, which was 500 miles due north of Syene. At noon, on the first day of summer, when he knew the sun was directly overhead in Syene, he measured the shadow of a tall pillar in Alexandria. He found the sun's rays made an angle of 7.5 degrees with the pillar. He reasoned, as you can see in the drawing, that if the sun's rays are parallel, then 500 miles is equivalent to 7.5 degrees of the earth's 360 degrees. Each 1,000 miles then is equivalent to 15 degrees. Since $360°/15° = 24$, the entire circumference of the earth must be 24,000 miles. Its diameter is 24,000 miles divided by pi. (To calculate the diameter of any circle, you divide its circumference by pi. (Pi is about 3.14.) Hence, the earth's diameter is about 8,000 miles; its radius 4,000 miles.

Early sailors knew that the altitude of the North Star decreased as they sailed south. Once they sailed south of the equator, Polaris dropped below the horizon. Such observations can be explained if the earth is a sphere. In fact, as you saw in the drawing on p. 3, the altitude of the North Star will be equal to the latitude from which it is viewed.

By the 18th century, educated people believed the sun, moon, and stars appeared to move across the sky because the earth rotated on its axis once each day. But direct evidence of the earth's rotation was not available before the middle of the 19th century, when Jean Foucault built his famous pendulum. A Foucault pendulum consists of a heavy metal bob suspended from a very long string. You may have seen one in a science museum.

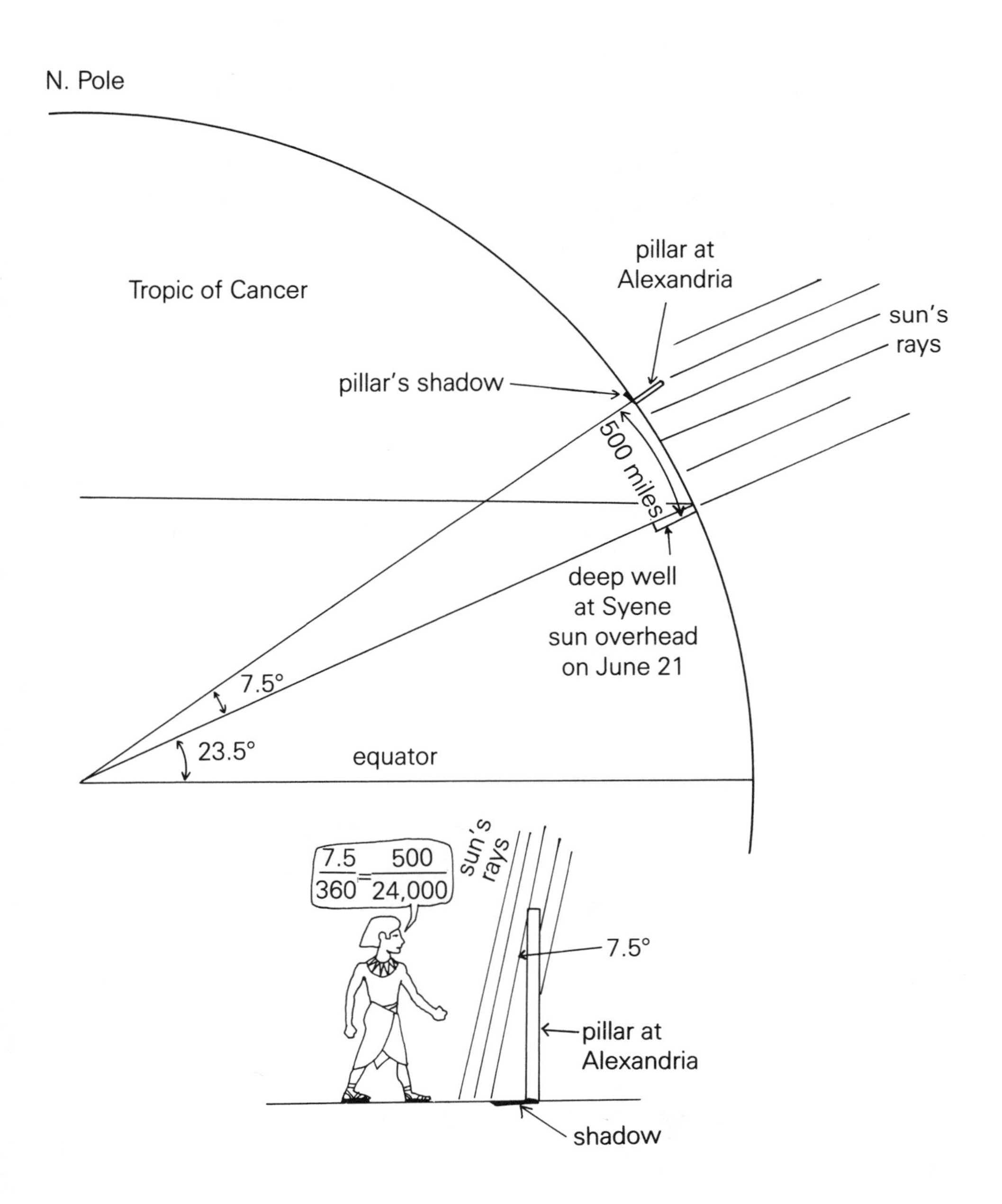
N. Pole
Tropic of Cancer
pillar at
Alexandria
sun's
rays
pillar's shadow
500 miles
deep well
at Syene
sun overhead
on June 21
7.5°
23.5°
equator
sun's
rays
7.5/360 = 500/24,000
7.5°
pillar at
Alexandria
shadow

Because a pendulum will maintain its direction of swing (Newton's First Law of Motion), Foucault knew, when he saw the pendulum change its direction of swing, that the earth must be rotating.

To make a model of a Foucault pendulum, hang a pendulum over a turntable. Pretend that the center of the turntable is the North Pole and set the pendulum swinging as in the drawing. You'll see

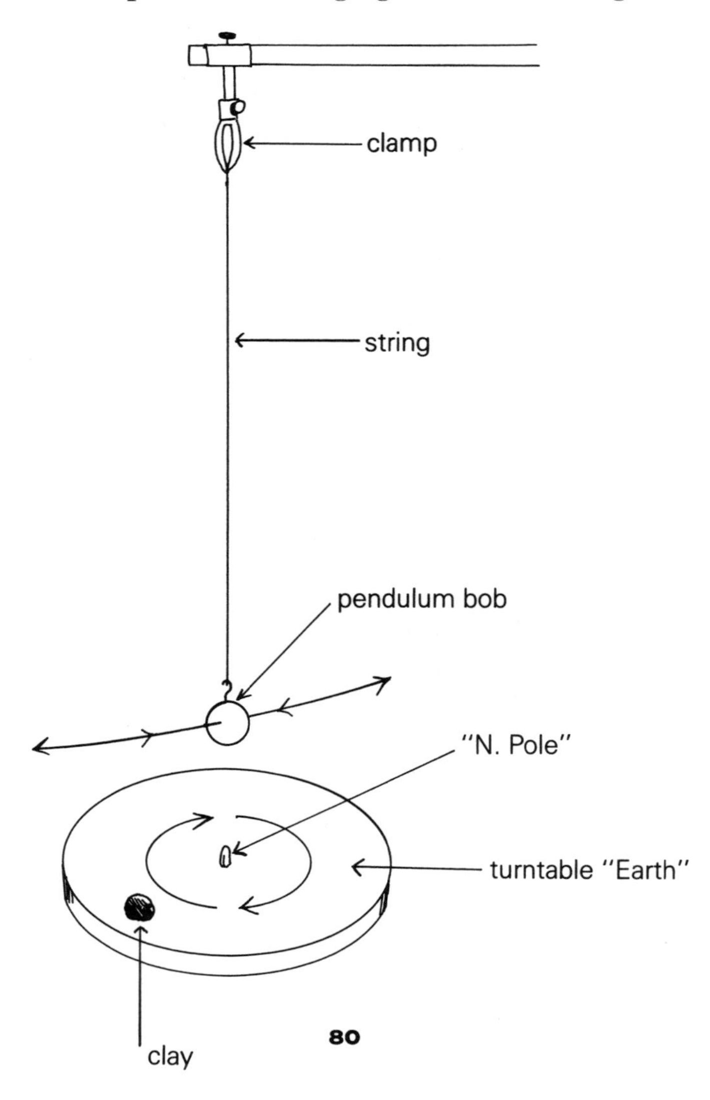

that the pendulum maintains the direction or plane of its swing. If a Foucault pendulum is swinging above the North Pole, it's direction of swing will appear to rotate through 360 degrees every 24 hours. To see why this happens, put a piece of clay on the turntable to represent you. Then slowly turn the turntable while the pendulum is swinging. As you can see, the direction of its swing appears to rotate relative to the clay.

## KEEPING SATELLITES ALOFT

When a satellite is launched, it acquires momentum from the rocket fuel used to increase its speed to that required for orbit. The path of a satellite is controlled by a complex guidance system. Radio signals sent from a control center activate sensors on the satellite that feed data into computers that automatically adjust the direction of the vehicle and keep it on a preplanned path to orbit. If you have a remote-controlled toy car or airplane, you're familiar with a simple guidance system.

Once a satellite is in orbit, no fuel is required to keep it there. You know from Newton's first law that a body in motion will maintain its motion unless a force acts on it. In the case of a satellite, there is a force—the force of gravity. The earth's pull on the satellite causes it to fall toward Earth just like any other body. But by the time it reaches orbit, the satellite has a speed of 5 miles per second (18,000 mph) along its orbit. The combination of its sideways speed and its acceleration toward the earth gives it a path that matches the earth's curved surface, as you saw in Chapter 3.

If the force on a satellite suddenly disappeared, what would its path be? To find out, put a marble on a level surface and put a rigid plastic cake cover over it. Gently swirl the cake cover so that the marble follows a circular path along the circumference of the cover. Once the marble "satellite" is in orbit, lift one side of the cake cover.

**HEOS was the European Space Research Organization's first interplanetary physics research satellite. It was launched by NASA in the late 1960s.**

What path does the satellite take? Does it continue to move in a circle? Does it fly outward? Or does it move in a straight line tangent to the circle?

Any object moving in a circular path is accelerating toward the center of the circle because there is a force pulling it inward. This inward force is called a *centripetal force.* In the case of a satellite, gravity provides the centripetal force. If you tie a ball to the end of a string and swing it around your head, you have to pull inward on the string to make the stone move in a circle. When a car goes around a curve, the tires push outward against the road. The road, in turn, pushes inward on the tires, supplying the centripetal force needed to make the car go in a circle.

If the road is covered with ice, there may be no centripetal force. As a result, the car will continue to move in a straight line that may carry it off the highway, just like the marble released from the cake cover.

## AN ACCELEROMETER

An accelerometer is a device that allows you to detect acceleration. If you have a small carpenter's level, you can use it as an accelerometer. If not, you can make one.

Find a tall, narrow plastic pill vial. Put a tiny piece of soap in the bottom of the vial. Then pour warm water into the vessel. Leave a little space at the top so a small bubble will be present after you close the vial.

Hold the vial in a horizontal, level position. Then accelerate the vial forward. You will see the bubble move in the direction of the acceleration. As the vial slows down and comes to rest, the bubble will move toward the rear of the vial, which is the direction of the acceleration when the vial decelerates.

## An Inward Acceleration

To convince yourself that there is an inward or centripetal acceleration on a body moving in a circle, use some clay to fasten an accelerometer to a turntable as shown. Be sure the bubble is in the center of the vial. What happens to the bubble when the turntable is spinning? What is the direction of the acceleration when the accelerometer moves in a circle? What happens to the acceleration when the turntable spins faster? When it spins slower? How can you tell?

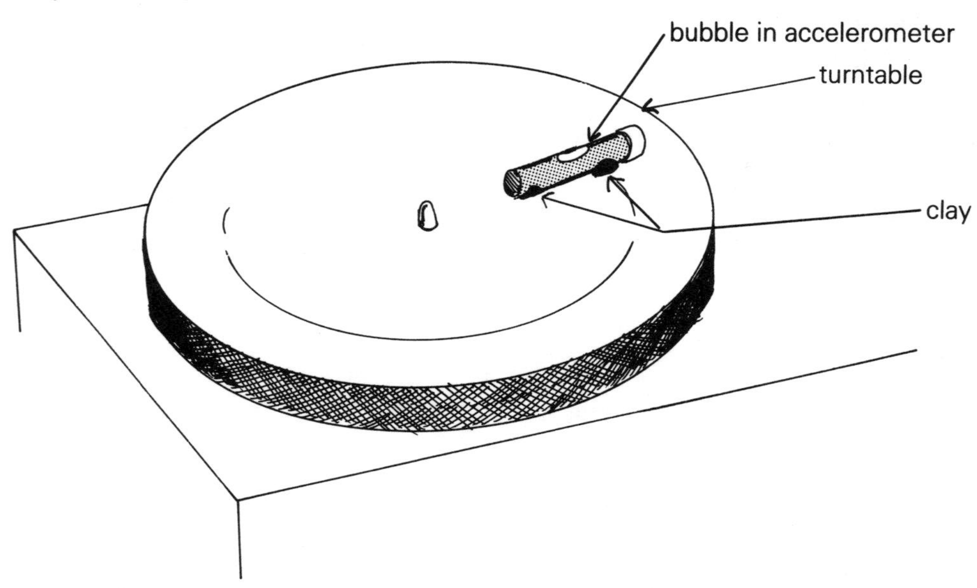


If you've ever played snap-the-whip, you know that you move faster as you get farther from the center of a circle. The same is true on a merry-go-round or a turntable. To see how the inward acceleration is related to the distance from the center, move the accelerometer closer to the center of the turntable. What do you find?

What do you predict the inward acceleration will be if you place the bubble right over the center of the turntable? Were you right?

## MEASURING CENTRIPETAL FORCES

To see how mass, radius, and period are related to the centripetal force needed to keep a satellite in orbit, you can build the apparatus shown in the drawing. You can use a rubber stopper to represent a satellite. Washers suspended from a string attached to the "satellite" will supply the centripetal force.

Take a ball-point pen apart. Remove the ink cartridge and use the barrel of the pen to swing the "satellite" as shown in the figure.

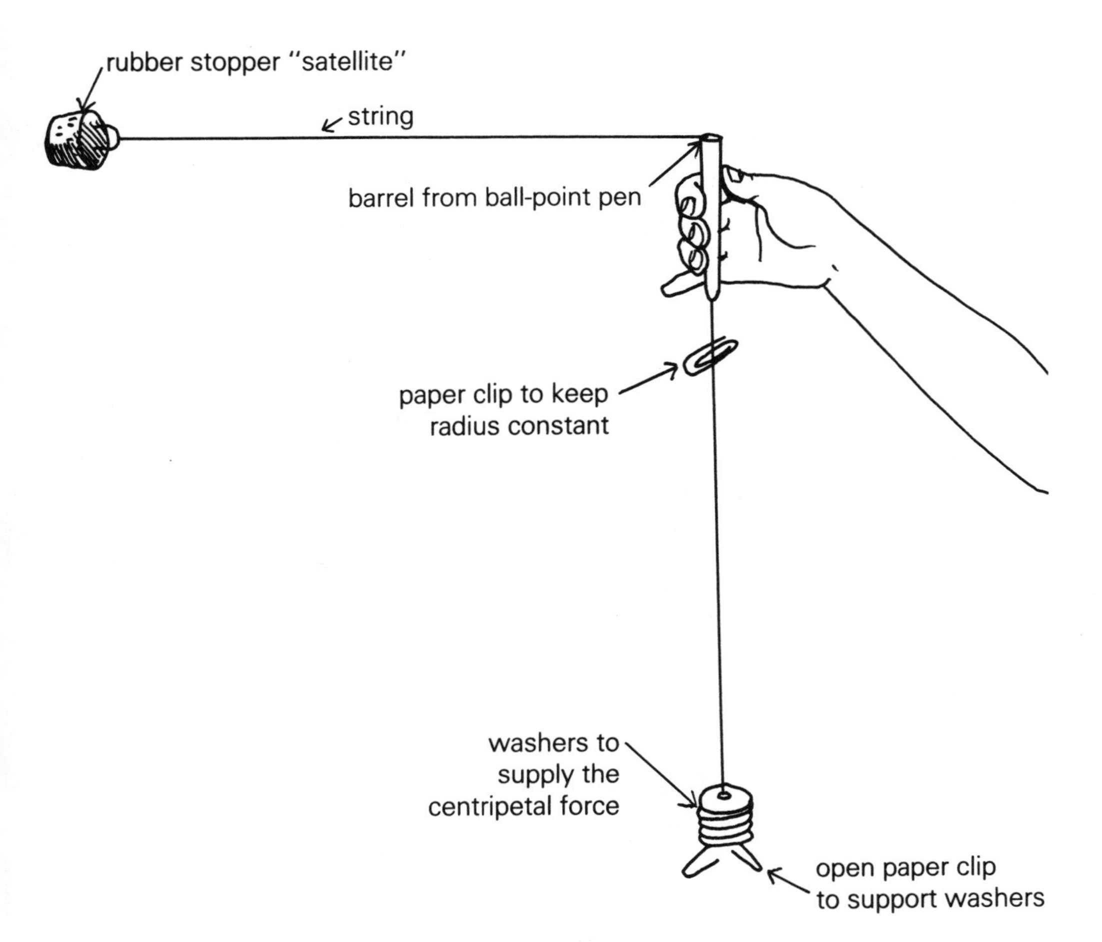

Thread a strong piece of thin nylon string, about 6 feet long, through the barrel of the pen. Use the smoother end of the barrel as the upper end around which the string will move in a circle. The lower end of the string can be tied to a paper clip that will support a number of identical metal washers. The weight of these washers provides the centripetal force needed to make the satellite move along a circular orbit.

Arrange the satellite so that its center is about 3 feet from the top of the pen barrel. Another paper clip can be attached to the string about an inch below the barrel. When you swing the satellite in its "orbit," keep this paper clip about an inch below the barrel so you can be sure that the radius of the orbit does not change. When you have the satellite moving smoothly at a radius of 3 feet, count the number of revolutions made by the satellite in 30 seconds. (A revolution is one time around the circular orbit.) The time required for the satellite to make one revolution is its *period*. You can find the period by dividing 30 seconds by the number of revolutions you counted. What is the period of your satellite? Why is it better to measure the time for many revolutions rather than just one?

What is the period of your satellite if you double the centipetal force but keep the radius the same? You can double the force by doubling the number of washers at the bottom of the string. How would you make the centripetal force four times as big? How much force is needed to halve the satellite's period?

If you halve the radius of the orbit, what force is needed to keep the period the same? One guess might be that at half the radius only half as much force would be required. Try it! Was it a good guess?

From Newton's second law, you might reason that if you doubled the mass of the satellite, you would need twice as much force to keep the satellite moving along the same orbit with the same period. Tape or tie another identical rubber stopper to the first one to double the mass of the satellite. With the same orbital radius (3 feet), does the period remain about the same when you double the centripetal force acting on twice the mass?

## Experiencing G-Forces

When you jump from a diving board into water, gravity causes you to accelerate at 32 feet per second per second. When you stand on the floor; the floor pushes up on you with a force equal to your weight. You feel what is called a one-G force. If you were accelerated along a track in a rocket sled at twice the acceleration that gravity provides, you would feel a two-G force on your back. In other words, you would feel as if your weight had suddenly doubled.

To experience at least a partial increase in weight, take a portable bathroom scale into an elevator. Stand on the scale and press the button for an upper floor. What happens to your weight, as recorded on the scale, when the elevator accelerates upward? What is your weight when the elevator moves at a steady speed? What happens to your weight as the elevator decelerates before coming to a stop? Can you predict what will happen to your weight when, after coming to a stop, you press the button for a lower floor?

Of course, you are no heavier or lighter in an elevator than anywhere else. But the forces causing you to accelerate make you feel heavier because you push back with an equal force. If the elevator accelerates downward, it doesn't push as hard on you; therefore, you can't push as hard on it, and so the force you exert on the scale resting on the elevator floor decreases.

The G-forces you experience in an elevator are quite small. If you ride a roller coaster or some of the other rides at an amusement park, you will experience larger G-forces. These forces will make you feel very heavy for short periods of time. In one of these rides, the force lasts longer. People stand against the wall of a huge "barrel" that can be made to spin. When it is spinning at full speed, the bottom of the barrel is lowered, but no one falls. The barrel is spinning so fast that the centripetal force exerted by the walls to keep the people going in a circle holds them firmly in place.

You'll find it interesting and fun to take your accelerometer with you to an amusement park. Are you able to predict the direction of the acceleration on these rides?

## DOING THE LOOP-THE-LOOP

In a loop-the-loop roller coaster, the cars stay on the tracks even when they are upside down. The centripetal force the tracks exert on the cars is greater than the force of gravity. Therefore, the inward acceleration of the cars (and the people in them) toward the center of the loop is greater than the acceleration due to gravity. As a result, the cars do not fall from the track. If the diameter of the loop is 40 feet, the cars only have to move at a speed of 18 mph at the top of the loop to stay on the track. Since the cars are moving much faster than that, there is no danger of their falling. In fact, when you are on the ride, you'll feel a force on the seat of your pants even when you are upside down.

You can see the same effect in this experiment to do outdoors. Place about a quart of water in a plastic pail and swing it in a vertical circle. Even when the moving pail is upside down, water will not spill. The bottom of the pail has a centripetal acceleration that is greater than the acceleration due to gravity. If you're in a bathing suit, slow down the pail to see when the water spills out.

## ON BOARD A SATELLITE

As you have seen with your rubber-stopper satellite, twice as much centripetal force is needed to keep a satellite with twice as much mass moving along the same orbit at the same speed. With real satellites this happens naturally. Doubling the mass of a satellite doubles its weight (the gravitational force exerted by the earth).

To double the rubber-stopper satellite's period along the same orbit, the force must be made one-quarter as strong. With real satellites, the force can be made smaller by placing the satellite farther from Earth. If the satellite is about 5.5 Earth radii above the earth's surface, its period will be 24 hours. At this altitude, the satellite has the same period as the earth. Do you see why communication satellites are placed at this altitude?

With real satellites, the force on the satellite is the force of gravity. As a result, the satellite and everything on it falls toward Earth at the same rate. If you were riding on a satellite, it would be similar to riding on a free-falling elevator, there would be no force on your feet. You would feel weightless.

You experience the sensation of weightlessness when you jump off a diving board into water. The feeling doesn't last long because you soon hit the water. You've probably experienced partial weightlessness on an elevator as it starts to descend. You can also experience weightlessness on a playground swing. If you're swinging high, you'll feel weightless at the peak of your swing. At the moment the swing stops climbing and starts its descent, both you and the swing will be falling at the same rate. You'll feel no pressure between your seat and the seat of the swing for a short time. As you descend along the circular path, you feel the swing pushing on you because it must provide the centripetal force needed to move you along a circular path.

During their training, astronauts experience longer periods of weightlessness by flying in airplanes that make long humplike arcs through the air. The planes fly at such a speed that the centripetal acceleration along their curved paths matches the acceleration due to gravity. As a result, the plane and all its contents are accelerating toward Earth at the same rate as any falling object. Sometimes, you'll experience the same effect when you go over an unexpected bump in a car.

# CHAPTER 5

# LIFE IN SPACE

Living in space is very different from living on Earth. Even the view is different. On Earth you see a blue sky and red sunsets. In space, the sky is black and the sun remains yellow as long as it is visible.

## BLUE SKIES AND RED SUNSETS

The sky appears blue because of the earth's atmosphere. The oxygen and nitrogen gases that make up the atmosphere absorb and emit light—a process called scattering. Because these gases tend to scatter the shorter wavelengths of light (blue) more than longer wavelengths (red), and because the scattered light is emitted in all directions, the sky appears blue. In space, where there is no atmosphere, light is not scattered. As a result, we can see light coming from stars, but the space around the stars appears black because there is nothing in the vacuum of space to scatter light.

On Earth, as the sun approaches the horizon, sunlight must travel through a greater length of atmosphere, as you can see from the drawing. Because sunlight encounters more and more gas as it sinks toward the horizon, more and more of the light at the blue end of the spectrum is scattered. The light that is least scattered is red. Hence, the setting sun appears redder and redder as it approaches the horizon because only reddish light comes through the air from the sun.

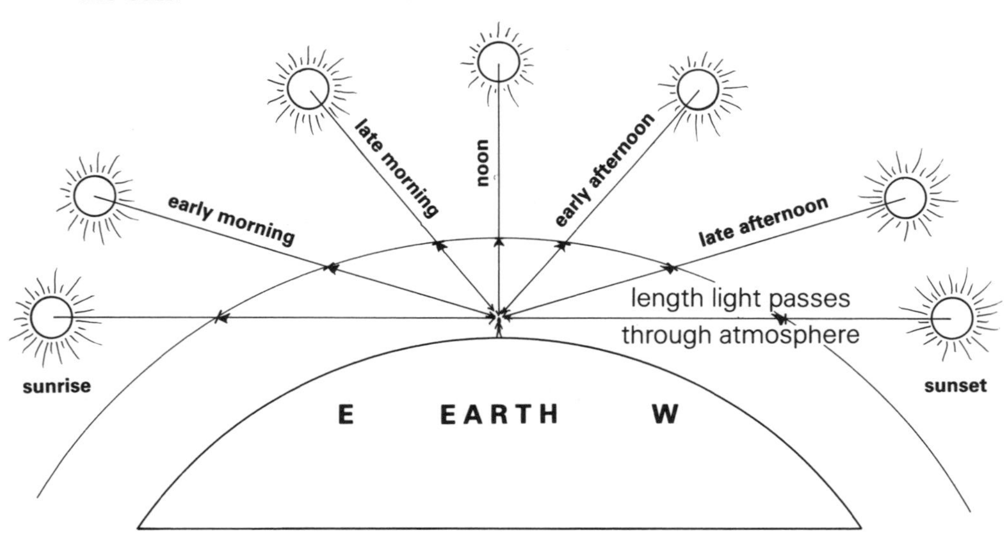


You can make a model sunset and see the blue "sky" caused by scattered light as well. In a dark room, shine the light from a slide projector through a fish tank full of water. To scatter light, add a small amount of powdered milk or a nondairy creamer to the water and stir. Notice how the water around the light begins to take on a bluish color. Continue adding small amounts of the powder and stirring. What happens to the color of the unscattered light coming through the water directly from the light?

## Life in a Weightless Environment

It's difficult to imagine the feeling of weightlessness. You can get a sense of what it would be like by floating in water. However, you can swim in water; you can't "swim" if you are stranded in the center of a space station. There's nothing to push against. If you were to find yourself on the ship's surface, you might try to walk. But when you pushed against the floor, you would move away from the floor (a direction we normally call "up") as well as forward. One way to avoid such unwanted motion is to wear shoes with Velcro soles and line the inside of the spaceship with the same material. That way, your feet will stick to the surface, and you won't find yourself bouncing off the walls.

What would you do if you were stranded and weightless in the center of a spaceship? This is but one of the many problems you might face in the weightless environment of space. The following activities will introduce you to a few of them.

## Pressure and Space

If you have ever dived deep into water, you know that the weight of the water produces a pressure that can exert forces in all directions. To see how the weight of water affects the pressure it can exert, use a small nail to punch three holes in the side of a Styrofoam coffee cup. Make holes near the bottom, top, and middle of the cup's side. Cover the holes with your fingers and fill the cup with water. Hold the cup over a sink and release your fingers. Notice how far the water in each stream is pushed outward. How is the pressure on the water coming out of each hole related to the weight of the water above the hole?

The pressure effect you have just seen is called *hydrostatic pressure*. The higher the column of water, the greater the pressure at the bottom of the column. The pressure increases with depth because the weight of a tall column of water is greater than that of a short column. But would there be hydrostatic pressure in a weightless environment?

To find out, fill the cup with water again. This time, after you remove your fingers from the holes, let the cup fall 6 or more feet into a tub of water. While falling, the cup and its contents will be accelerating at the same rate. It's similar to a weightless environment.

What do you notice about the water streams while the cup is falling? Try it several times to be sure. Would there be hydrostatic pressure on a spaceship?

To create pressure in a spaceship, you would have to provide some force other than weight to move fluids. To see one way this might be done, fill a large, strong balloon with water. Seal the neck of the balloon with a tie band. Use a small pin to make a tiny hole on one side of the balloon. A stream of water, similar to the one you found coming from the side of the coffee cup, will emerge. Now drop the balloon into a tub of water as you did the coffee cup. Repeat the experiment several times to be sure. Does the stream stop flowing this time?

What's one way you could provide the pressure needed to move fluids in a spaceship?

## Dining in Space

Eating in space presents problems. Food won't stay on your plate. If you nudge it with a fork, it'll keep moving. Milk won't pour from a glass—it's weightless. You might wonder if you could eat at all in a weightless environment. After all, there's no force to carry the food from mouth to stomach.

Is starvation a natural consequence of weightlessness? To find out, fill a paper cup with water, rest it on the floor, and place a straw in it. Lie on a chair and bend your head down over the water so that your stomach is higher than your mouth. Can you overcome gravity and drink water through the straw when you are in this position? Could you drink and eat if you were weightless?

Astronauts living in the weightlessness of space drink from plastic bottles. Squeezing the bottles with their hands provides the pressure needed to move the liquids into their mouths. Some of the food they eat is also squeezed from plastic bags. The involuntary contraction of muscles in the esophagus moves food and liquids from mouth to stomach.

Sticky foods can be served on plates fastened to a table, but care must be taken not to move spoons and forks too fast or the food may keep on moving when the silverware stops. Have you ever seen astronauts with egg on their faces because they forgot the first law of motion?

Imagine preparing and eating a meal in a spaceship. Compare such a meal with one on Earth. List all the precautions you would have to take and all the things you would have to do differently if you were eating in space.

You can eat and drink in space, but the food doesn't have much taste. In weightless conditions, body fluids do not tend to drain into your legs. As a result, your face appears fatter and you have a runny nose and more fluid in your chest—all the symptoms of a cold, the same symptoms that make it difficult to taste and smell.

## Convection and Space

You can eat in space, but not by candlelight. When a candle burns on Earth, the hot gases produced from the burning wax rise as cooler, denser air moves in around the candle, forcing the hot, lighter gases upward.

A hot gas weighs less than an equal volume of the same gas when it is cool. This is true of liquids as well. Weigh 100 ml of water. Then weigh an equal volume of cooking oil. Which is heavier? Now mix some cooking oil and water in a cup. Which liquid rises to the top?

Fill a medicine vial with cold water. To a second vial add several drops of food coloring. Then fill the second vial with hot tap water. Using an eyedropper, remove some of the colored hot water. Carefully lower the eyedropper into the cold water, and very gently squeeze out a drop of the colored hot water. Does hot tap water rise or sink in cold water?

Repeat the experiment, but this time color the cold water and gently squeeze it into some clear hot water. Does the cold water rise or sink in the hot water?

In a weightless environment, there is no force to move carbon dioxide and other waste gases from a burning candle. As a result, fresh, cooler, denser, oxygen laden air will not move in to replace the waste gases. The candle is quickly extinguished by its own carbon dioxide. No fire extinguisher is needed.

## A Weightless Burning Candle

**Ask an adult to help you with this experiment. It involves dropping a burning candle.**

Place a birthday cake candle in a small lump of clay on the bottom of a large, wide-mouth glass jar. **Have an adult light the candle.** Screw the cover onto the jar and measure how long the candle will burn inside the closed jar.

Remove the cover and candle, invert the jar, and move it up and down to get out any gases formed by combustion. When you are sure the jar is filled with fresh air, **have an adult relight the candle** and cover the jar at the top of a step ladder on a lawn. The jar can then be dropped onto some soft pillows or a pile of sand or soft earth. Stand back away from the falling jar and watch the burning candle as it falls. Does the candle continue to burn, or does it go out?

In doing this experiment, both you and your adult helper **should wear protective shielding over your face in case the jar should break. It would also be a good idea to have a fire extinguisher nearby.** It is not likely that breakage will occur if the jar falls onto soft pillows, but it is wise to be overly cautious where fire and glass are concerned.

On Earth, differences in density (weight per volume) resulting from expansions or contractions caused by temperature changes create convection currents. In the weightless environment of a spaceship, nothing sinks or floats, so there are no convection currents. To be sure there is fresh air for the astronauts to breathe, a circulating system must constantly force air around the spaceship.

When crystals are grown on Earth, convection currents created by the heat released in the process often produce flaws in the crystals. Companies that grow crystals, such as silicon crystals, which are used in computers, hope they will be able to produce better crystals with greater efficiency in space factories where there are no convection currents.

## SEDIMENTATION

Put some mud, sand, and water into a plastic vial. Cover the vial and shake it. Watch the solid matter settle. Which particles seem to settle fastest? Slowest?

The settling of solid particles to the bottom of a liquid is called *sedimentation* — a process commonly used in chemistry to separate mixtures. Would sedimentation occur in a weightless environment?

Repeat the experiment, but this time add sawdust to the mud, water, and sand in the vial. What happens this time after the vial is shaken? Can you explain why some material moves to the top and some to the bottom of the vial?

Chemical reactions that take place between a liquid and substances that are not soluble in the liquid would occur much faster in a space laboratory orbiting Earth. Can you explain why?

## GROWING PLANTS IN SPACE

For plants to grow on Earth, water, carrying dissolved minerals from the soil, must move upward through the roots and stems to the leaves where plants manufacture food. Some believe that plants will grow faster in a weightless environment because the water can move faster along the stems if gravity is not present to retard it.

To check up on this idea, dip a piece of paper towel into some water to which a few drops of food coloring have been added. Notice how water "climbs" up the towel. If you watch this under a microscope, you'll see that the water fills the tiny spaces between the wood fibers that make up the towel. There are similar tiny channels in the stems of plants through which water moves. Water is attracted to the walls of these narrow pathways. Because water is cohesive (holds together well), water that is attracted to these walls pulls more water along with it, filling the channels. The movement of liquids upward in small spaces is called *capillary action.*

To see how gravity affects capillary action, set up the experiment shown in the drawing. Cut two identical strips from a paper towel. Hang one vertically. Lay the other one out horizontally. One end of each strip should rest in a small dish of colored water. Watch the colored water move along the two strips. How far has it moved along each strip after 10 minutes? After 20 minutes? After an hour? In which strip does water move faster? Does the movement of the water ever stop in either strip?

You might repeat this experiment using celery stalks in place of paper towel strips. But you may have to cut across the stalks in or-

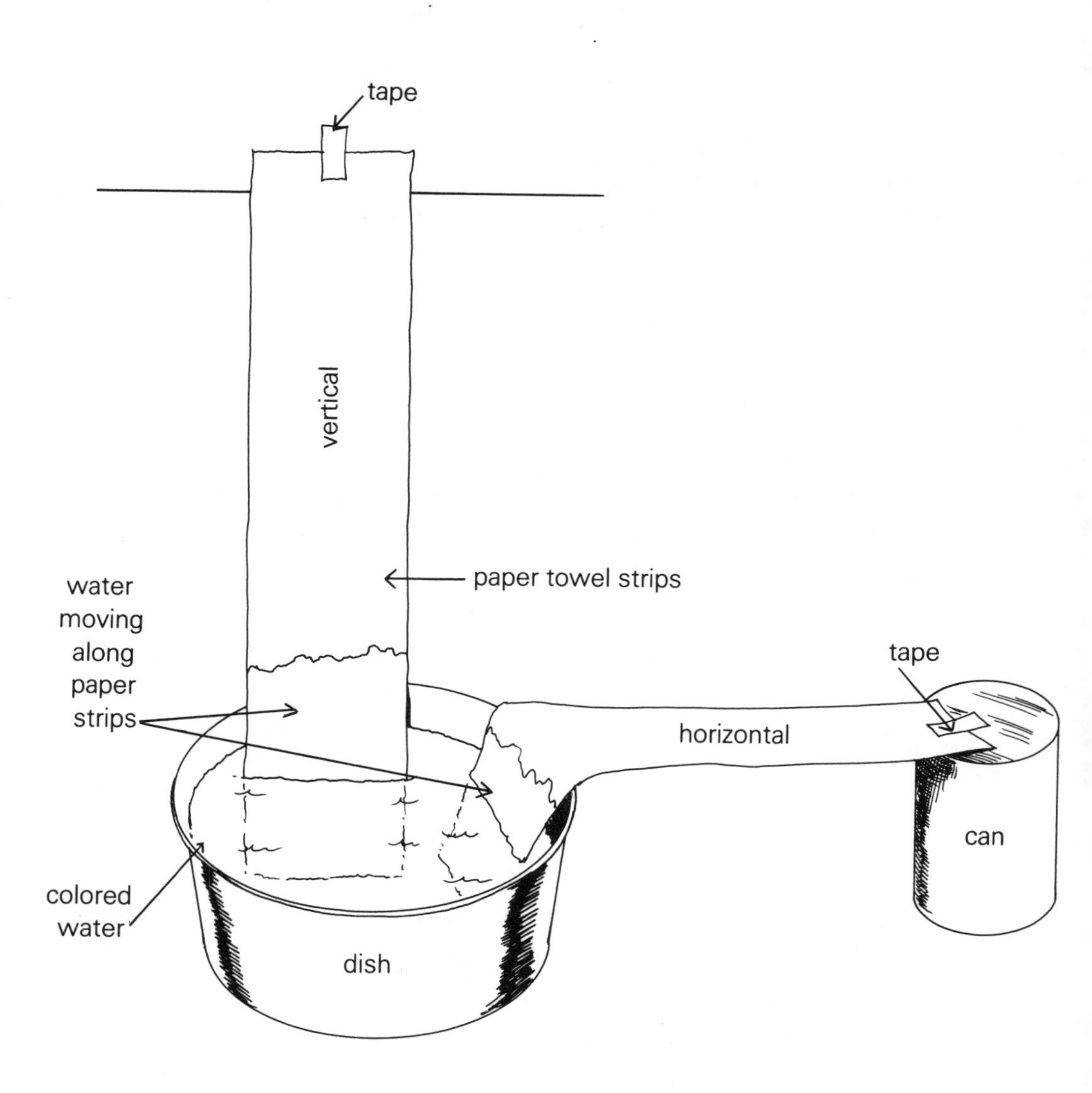


der to see how far the water has moved.

Do you think water would move faster through plants in a weightless spaceship than through plants growing on Earth?

## A Bathroom in Space

Without gravity how would you take a shower? A pressured tank could be used to force the water into a stream, but once it hit your body, the droplets would bounce all over the ship.

One way to avoid this problem is to enclose the shower from top to bottom so no water can escape. At the bottom of the shower, low pressure is created in a drain by means of a vacuum pump.

Another way to create low pressure is with fast-moving streams of air. To see how this works, hang two balloons from the top of a doorway so they are a few inches apart. Now blow air between the balloons. What happens to the balloons?

If you examine an atomizer, such as the kind used to spray cologne, you'll see that it works in a similar way. An airstream is directed across the top of a tube that dips into the liquid. This reduces the pressure at the top of the tube and so cologne is forced up the tube by the higher pressure over the liquid.

In a spaceship, fast-flowing air and water streams under the toilet create a low pressure that causes material dropped into the toilet to move into the drains leading to collecting tanks where waste is stored.

## Muscles in Space

On Earth, the stresses created by the weight of your flesh as it pulls on the bones of your skeleton in some way cause new bone matter to form, replacing any that may disappear through wear and metabolism. But in the weightlessness of space, this does not happen unless an effort is made to maintain muscle tone. Early ventures into space led space scientists to realize that astronauts would have to engage in vigorous daily exercise.

In a zero-G environment, your bones respond in the same way as the bones of bed-ridden people on Earth; they lose calcium and become weaker. Through lack of work, your muscles (including heart muscle) and blood vessels lose their tone. Blood volume, red blood cells, and the salt concentration in your body fluids all diminish. As a result, Charles Conrad, Joseph Kerwin, and Paul Weitz, the first crew aboard Skylab, who spent 28 days in space in 1973, were barely able to walk when they returned to Earth. They felt faint because blood tended to pool in their lower body, their heart rates were higher than normal, they had lost weight, and their muscles ached from having to carry their new weight.

It was found that these effects could be greatly reduced by 90 minutes of vigorous exercise on an ergometer each day. In addition to daily, vigorous exercise, Russian cosmonauts wear suits with crisscrossing elastic cords. Every movement requires muscular effort. It's uncomfortable, but it helps them maintain muscle tone during lengthy stays in the Soviet space station.

## Finding Your Weight in Space

During their time in space, astronauts keep records of what they eat, how they feel, their height and weight, and other statistics. But how can they determine their weights in a weightless environment?

Actually, they don't measure their weight; they measure their mass, which is the same everywhere. Since they are weightless, they will not stretch the springs of any scale used to measure weight. Even the equal arm balances used to measure mass in laboratories cannot be used here. Such balances depend on the equal gravitational pull on the masses at either end of the balance. However, there are balances, called inertial balances, that can be used to measure mass anywhere. Such balances do not depend on gravity for their operation.

The inertial balances used in space stations or space laboratories are large enough for a person to sit in. But you can make and calibrate a small model inertial balance quite easily. Use a C-clamp to fasten a yardstick to the side of a bookshelf as shown in the drawing. Fasten a 100 gram mass to the end of the yardstick with tape. Then measure the time it takes the yardstick to swing back and forth (oscillate) fifty times. Repeat the process again using heavier masses. Each time, record the mass and the time required for the yardstick to make 50 oscillations. Does the oscillation time of the yardstick depend on the mass attached to the end of the stick?

Plot a graph of the time to make 50 swings versus the mass attached to the yardstick. Plot time on the vertical axis and mass on the horizontal axis. Connect the points plotted with a smooth line.

Now attach an object of unknown mass to the end of the yardstick. You might use another C-clamp or a stone. Measure the time it takes for the unknown mass to make 50 oscillations. Then use the graph you have made to estimate its mass. If you have a regular balance, you can check the mass of the unknown and see how close your prediction was.

To see that gravity does not enter into this measurement of mass, hang one of the masses you used from a long string so that it does not rest on the yardstick, as shown in the drawing. Tape it to the yardstick so it will move when the yardstick moves. The mass will now move with the yardstick balance, but it is not exerting any gravitational force on the yardstick. How does the time required to make a certain number of swings now compare with the time it took to make the same number of swings when the same mass was pulling downward on the yardstick?

## Energy in Space

People living in space have the same energy needs as you. They must stay warm, they need light when they are in the earth's

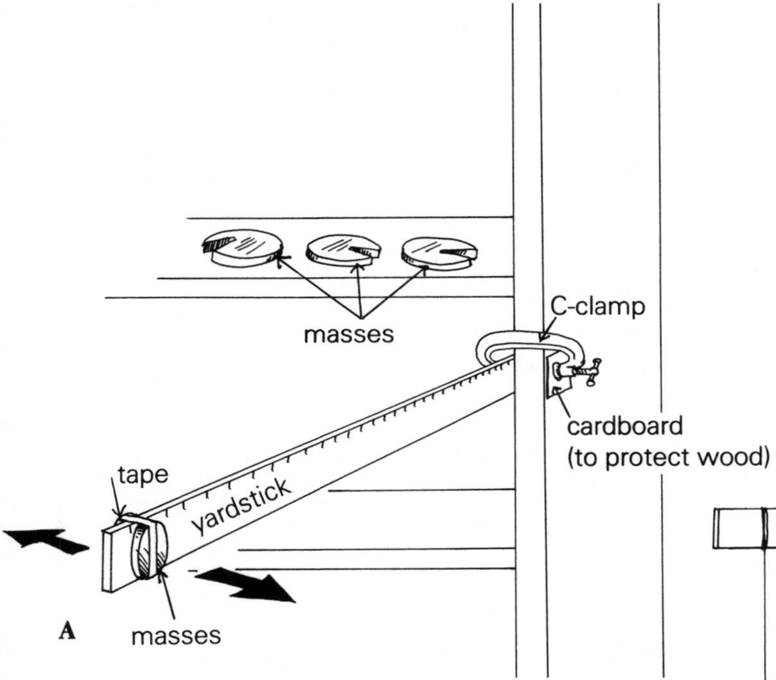

masses
C-clamp
cardboard
(to protect wood)
tape
yardstick
A
masses

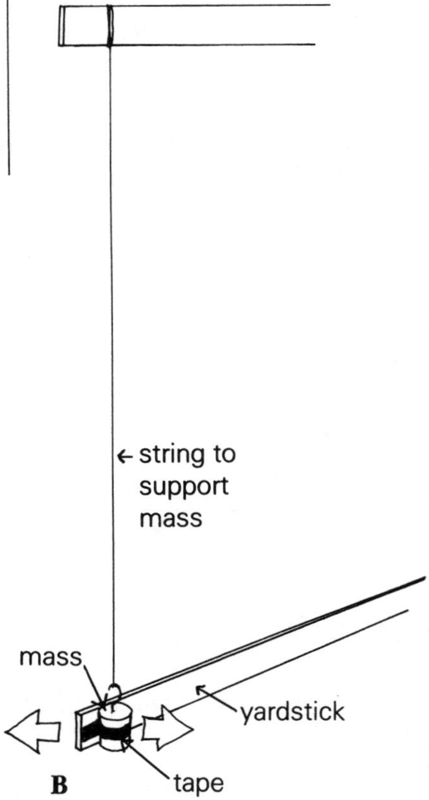

string to
support
mass
mass
yardstick
B
tape

shadow, and they need electricity to run motors, stoves, air conditioning, etc. The energy required in spaceships comes from the sun. Giant panels of photovoltaic cells convert sunlight into electricity.

If you can get some photovoltaic cells, you might connect them to a small electric motor and see if you can make the motor run with solar power.

Solar energy can also be used to heat air and water in a spaceship. In fact, sometimes the heat produced can raise temperatures so high that it becomes uncomfortable. Then the problem becomes one of getting rid of excess solar energy.

To see how color affects the conversion of sunlight to heat, paint the inside surface of an aluminum pie pan with flat black paint. Place this pan side by side with one that has not been painted on a sheet of cardboard in a warm, sunny place. Pour a cup of water in each pan and leave them in direct sunlight for several hours. Every few minutes use a thermometer to measure the temperature of the water in each pan. In which pan does the water get hotter? Can you explain why?

Repeat the experiment with two more aluminum pans. Add black ink to the water in one pan. Which pan do you predict will get hotter?

You might also place a series of pans side by side and place equal amounts of water in each. Color the water in each pan differently using food coloring. Does the color of the water affect its ability to absorb heat?

Repeat the experiment once more. This time paint both pans black. Cover one pan with aluminum foil. Leave the other uncovered. In which pan does the water get hotter?

Based on your experiments, what color containers would you use to absorb solar energy? How would you reflect away unwanted sunlight?

As you have seen, life in space can be quite different from life on Earth. The challenge for space scientists and engineers is to design space homes where people can live normal, healthy lives. The problems of weightlessness must somehow be overcome if people are to spend their lives working in space. We'll look at some ways that can be done in the next chapter.

# CHAPTER 6

# OUR FUTURE IN SPACE

We know how to send humans into nearby space, but is it practical and profitable to do so? Much of the information that can be gathered from satellites can be done without humans on board. Weather, communication, LANDSAT, and other data-gathering satellites can operate perfectly well without humans in space.

As Earth's resources become more and more scarce, their cost will grow to the point where such projects as mining the moon or asteroids will become economical. In the meantime, we must be prepared to act when Earth's resources near depletion. We must be ready to find, extract, and process the materials we need in space. Indeed, one way to conserve Earth's present resources is to send people into space now. Such expeditions would not only reduce pressures from the growing world population, but provide alternative energy sources that will lessen the rate at which we use our limited energy sources on Earth.

## Preparing to Live in Space

When the resources found in space become essential to our survival, we must be prepared to use them. And there must be suitable living accommodations for those who will work in space. Most people will not spend several hours a day engaged in physical exercise to retain muscle and blood vessel tone. We must find a practical way to create an artificial gravity and thus avoid the bad effects of weightlessness on the human body. Research in preparing for a practical living environment in space can be coupled with other research where space offers the best environment. Where else can scientists find a better vacuum, colder temperatures, fewer germs, or a better weightless environment?

## Industry in Space

Many believe that some industries could produce better products at less cost if they were located in space right now. Zero gravity makes it possible to manufacture perfectly spherical ball bearings. In a weightless environment, the weight of the metal cannot overwhelm the forces of surface tension that hold the melted metal together in perfect roundness. A lack of convection currents makes it possible to prepare uniformly mixed alloys and superconducting materials as well as perfect semiconductor crystals and the silicon chips essential to computers.

Devices requiring a near-perfect vacuum could be built in space, where there would be no need to pump out air. The device could be built around the natural vacuum in space. The sterile vacuum above our atmosphere would be the ideal place to prepare materials that must be germ-free. And tests aboard spacecraft have shown that the preparation of many drugs can be done more efficiently in space.

One of the major industries in space will be the production of electricity from solar energy. In space, there are no clouds or atmosphere to absorb or reflect the sun's radiation. Vast arrays of photovoltaic cells will convert sunlight to microwaves that can be beamed down to earth. Because microwaves can pass through clouds, they can be transmitted constantly from space to Earth. On Earth, receiving stations will convert the microwaves to electrical energy and send it out over the national power grid.

## MINES ABOVE EARTH

Sending the material needed to build solar power stations into space would be very expensive and uneconomical today. But if we built the power stations from materials mined on the moon, the huge costs involved in lifting the materials into space would be eliminated. Moon rocks contain ores rich in aluminum, magnesium, titanium, and iron, all of which could be used to manufacture solar power stations in orbits about Earth. The ores also contain oxygen, which the miners would need for breathing.

Initial costs would be high because the materials needed to build mines on the moon would have to be lifted from Earth. But once these mines were established, the costs involved in building the power stations would be much less than if they were built from Earth's resources. This is because it requires only 5 percent as much energy to lift matter off the moon as it does to lift it off the earth. The ore mined on the moon could literally be thrown into space by a machine on the moon that could accelerate the rocks to escape velocity. Away from the moon, on a "space island," the ore would be "caught" and processed. The metals and alloys produced on the

**A** photograph of the sun taken from Skylab 4 on December 19, 1973. Note the spectacular solar flare nearly 370,000 miles in width.

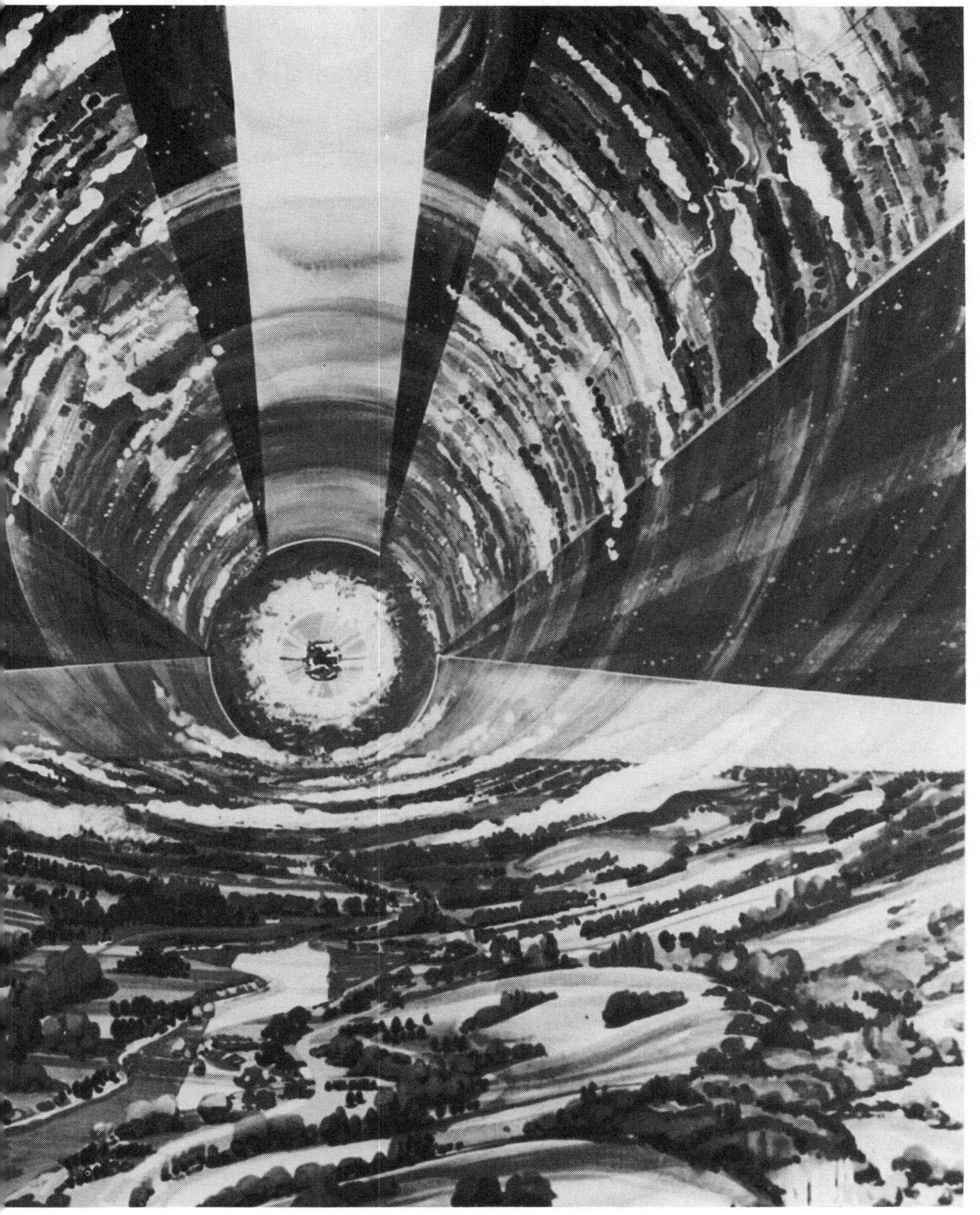

Inside a future space colony.

island would then be used to build the solar power stations that would beam microwaves back to Earth.

At first, the island in space would be relatively small to avoid the large costs involved in bringing materials from Earth. But after sufficient metals had been manufactured, a large, spoked, doughnut-shaped structure about a mile in diameter would be built around the factory to allow workers and their families to live comfortably in space. The outer surface of the doughnut would be covered with the waste slag produced as metals are extracted from lunar ores. The slag would protect the colony from cosmic rays, dangerous radiation, and meteorites. Mirrors that can be rotated will reflect light into or away from the colony to create night and day. The wheel will turn about once a minute producing an artificial gravity. To people living in the colony, "down" would be the outside surface of the doughnut. That's where they would feel a force on their feet. In response to the inward force that the doughnut's surface would exert on the people, the people would push back with an equal but opposite force. To them it would feel like Earth, where they pushed against the ground with their weight and the ground pushed back with an equal force.

All the food needed would be grown on the space island in soil transported from the moon. Growing food would eliminate the cost involved in transporting it from Earth, and the green plants would produce oxygen that the inhabitants needed for breathing. Chemicals needed to grow specific plants would be added to the soil. Additional oxygen would be obtained from the processing of lunar ores, but most of this oxygen would be used to burn the hydrogen used to fuel space shuttles and other means of transportation.

All waste material in the colony would be recycled. The island would initially depend on the earth for basic chemicals, particularly organic compounds that contain carbon. Eventually, as more and larger islands are built, spaceships, perhaps powered by solar sails, would travel to the asteroids. Space sails are extremely thin sheets of aluminum foil that would be spread out into an area miles across. The solar wind, particles of matter and light emitted by the sun, striking the giant low-mass sail would provide the energy needed to move the sail spaceship. Ships like these, or ships powered by hydrogen fusion, might travel to the asteroids, where compounds rich in carbon, nitrogen, and hydrogen as well as water could be tugged back to space factories for processing.

## Into Outer Space

Once life in space became common and Earth's orbit was no longer a frontier, humans would seek to conquer the next frontier. Journeys to other stars and solar systems would be planned. And someday, perhaps a century from now, a group of new pioneers would set sail for the stars.

Since the nearest star is over 4 light years away, these pioneers might never see Earth again. Even if they traveled at one-tenth the speed of light, which is more than ten thousand times faster than present day spaceships, it would take 50 years to reach even the nearest stars. (Astronomers suspect that stars with planets these space explorers might settle are at least 10 light years away.) Their spaceship would have to be a world of its own. If advances in medical science increased the normal human life span to a thousand years or more, a journey of several hundred years might be viewed as but one of several careers in a future human's life.

Because of the costs involved in building and moving a spaceship that could travel to the stars, its size would have to be limited, and so would the population on board. A small population would eliminate the advantage of professional specialization that our society enjoys. There will be a doctor who will be a general practitioner in the broadest sense. In addition to being surgeon, pediatrician, neurologist, and all other specialists, it's likely that he or she would serve also as the veterinarian. The plumber would have to be the electrician and carpenter as well. A teacher would probably have to be familiar with a number of subject areas and grade levels.

Could such a group of unspecialized individuals on so sophisticated a project be successful? Would the population outgrow its space? How would a second generation be trained? Would these space travelers meet other intelligent life from other planets such as ours? If they did, how would they communicate? These are questions for which we have no answers today. But in view of the pioneer spirit of humans, they will probably be explored, and perhaps answered over the next several hundred years.

The problem of specialization may not be as difficult as you might think. The skills needed to build, analyze, and educate might be stored in machines—computers and robots. The people chosen for the mission would be troubleshooters, people who maintain and use the computers and robots. Information would be stored in computers. Manufacturing skills would be programmed into robots. The spaceship doctor would present a set of symptoms to a computer, and the computer would respond with a diagnosis or ask additional questions that the doctor would answer. Should unfamiliar surgery or tasks be required of the doctor, he could refer to a detailed, illustrated description of the operation or procedure stored in the computer's memory.

Crew members on these spaceships would have to know how to learn, how to get information quickly, and how to use it to solve problems that could not be anticipated. The success of these missions would depend, as success always has, on the ingenuity of those who made decisions. And the decisions that will be made could not be obtained by radio contact with Earth. After the ship was one light year into its voyage, it would take more than two years

Jupiter as seen from Voyager 1 on February 5, 1979. From this planet, future starships may obtain the hydrogen needed to fuel the fusion reactors that will propel them beyond the solar system.

to get a response to any question transmitted to Earth, because the maximum speed of radiation of any kind is the speed of light.

Routine procedures would be done by robots, routine analyses and the solution of most problems would be carried out by computers. The education of the younger members of the crew and those born on the journey would be performed largely by computers, on an individualized basis. "Socratic" computers, blessed with infinite patience, would be able to talk to students, ask them questions, analyze their responses, and present new questions that would lead students to a thorough and insightful understanding of the subject matter being studied.

In addition to maintaining and operating the machinery, the humans on board would be free to deal with those questions or procedures not stored in the memory of the machines. What those questions and procedures would be we do not know today. A hundred years from now, sophisticated computers will do things that we would not dream of today. However, many people believe that there are tasks and problems that machines cannot do or be programmed to anticipate. In any case, all the machinery in the universe would not keep humans from wanting to see and explore whatever lies beyond the latest pioneer outpost.

## NEW CIVILIZATIONS

Would the spaceships that traveled to outer space carry the founders of new civilizations, or would these explorers maintain their nomadic ways, seeking ever new paths through the galaxy and beyond? We don't know, of course. But it's quite possible that these pioneers would do both. Settlements probably would be made on planets that have the necessities for human life—water, oxygen, and other chemicals essential for life—while some members of the spaceship continued the search for other possible sites. After all, those born during the journey would have never experienced life on a planet. For them, settling on such a small part of the universe might seem confining. They might prefer to search for other inhabitable planets or for other forms of life that could communicate with them. By then, our search for extraterrestrial intelligence might have borne fruit. Perhaps we would have received radio, laser light, TV, or some other signals from another civilization and will know where to search.

Eventually, human outposts might be established throughout the galaxy and, perhaps, beyond. But you might not recognize these pioneers. The human species, homo sapiens, has evolved into its present form on Earth over the last million years. Biologists have seen that members of a species that become separated and isolated evolve along diverging paths. Hence, it is likely that humans who became isolated from Earth homo sapiens would evolve in different ways. Aside by language differences, which would not surprise us, we would probably not recognize the features of these future humans as belonging to members of our species.

We expect human evolution to follow different paths on different planets or spaceships across the galaxy, and the growth of genetic engineering techniques would have its effect too. If we can transfer and alter the genes that control the development of our bodies, then it seems likely that those genes that best adapt humans to their new planets will be the ones chosen. Such techniques would probably reduce the time necessary for a species to be significantly altered from a million years to a few hundred or less. Over centuries, because of differences in environment, the choice of genes selected would vary from one civilization to the next. If our descendants met their long lost cousins, not only would they fail to see familiar family features, they might not even recognize the species.

Unless major changes in the laws of science take place, and that is always a possibility, the intergalactic wars presented on movie screens would never become a reality. Even though galaxies are in clusters, they are still millions of light years apart. To travel from one galaxy to the next, even at the speed of light, would take millions of years. By the time the space warriors reached their foes, their descendants either would have forgotten what the fray was about or lost interest. Fortunately, star wars will remain in the realm of fiction. We can hope that future journeys into space will be peaceful and fruitful.

**A** rising Earth as seen by astronauts on the moon. Moon miners will enjoy a similar view of Earth every month.

# APPENDIX

## SCIENCE SUPPLY COMPANIES

- Carolina Biological Supply Co.
  2700 York Road
  Burlington, NC 27215

- Delta Education
  P.O. Box M
  Nashua, NH 03061

- Edmund Scientific
  101 East Gloucester Pike
  Barrington, NJ 08007

- Nasco Science
  901 Janesville Road
  Fort Atkinson, WI 53538

- Schoolmasters Science
  P.O. Box 1941
  Ann Arbor, MI 48106

- Ward's Natural Science
  P.O. Box 1712
  Rochester, NY 14603

# BIBLIOGRAPHY


Apfel, Necia H. *Astronomy and Planetology.* New York: Franklin Watts, 1983.

Gardner, Robert. *Energy Projects for Young Scientists.* New York: Franklin Watts, 1987.

__________ *Ideas for Science Projects.* New York: Franklin Watts, 1987.

__________ *Space: Frontier of the Future.* New York: Doubleday, 1980.

Greenleaf, Peter. *Experiments in Space Science.* New York: Arco, 1981.

Hewitt, Paul G. *Conceptual Physics: A High School Physics Program.* Reading, Massachusetts: Addison Wesley, 1987.

Irvine, M. *Satellites and Computers.* New York: Franklin Watts, 1984.

Kaufmann, William J. III. *Universe.* New York: W.H. Freeman, 1985.

Kutter, Siegfried G. *The Universe and Life.* Boston: Jones and Bartlett, 1987.

Levitt, I. M. and Marshall, Roy K. *Star Maps for Beginners.* New York: Simon and Schuster, 1964.

McKay, David W. and Smith, Bruce G. *Space Science Projects for Young Scientists.* New York: Franklin Watts, 1986.

Moore, Patrick. *Exploring the Night Sky with Binoculars.* New York: Cambridge University Press, 1986.

__________ *Naked-Eye Astronomy.* New York: W. W. Norton & Co., 1965.

__________ *The New Atlas of the Universe.* New York: Crown Publishers, 1984.

Vogt, Gregory. *The Space Shuttle: Projects for Young Scientists.* New York: Franklin Watts, 1984.

# INDEX

# ABOUT THE AUTHOR

Robert Gardner is chairman of the science department at Salisbury School, Salisbury, Connecticut. He has taught at a number of National Science Foundation teacher institutes, including one in Ajmer, Indian, and was a staff member of the elementary science study of the Education Development Center. Mr. Gardner is the author of a number of books for young people, including *Kitchen Chemistry, The Whale Watchers' Guide, Science around the House*, and with David Webster, *Science in Your Backyard.*